American Mathematical Society

TRANSLATIONS

Series 2 • Volume 200

Advances in the Mathematical Sciences—48

(*Formerly Advances in Soviet Mathematics*)

# Methods of Qualitative Theory of Differential Equations and Related Topics

## Supplement

L. Lerman
G. Polotovskiĭ
L. Shilnikov
Editors

**American Mathematical Society**
Providence, Rhode Island

Translation edited by A. B. Sossinski

2000 *Mathematics Subject Classification.* Primary 34Cxx, 37Cxx; Secondary 14H99.

ABSTRACT. This book contains papers written by collaborators and students of E. A. Leontovich-Andronova. The papers are mainly devoted to new results in qualitative theory of ordinary differential equations and in the theory of smooth dynamical systems.

This volume is of interest to researchers and graduate students working in differential equations, dynamical systems, and related areas.

---

Library of Congress Card Number 91-640741
ISBN 0-8218-2663-8
ISSN 0065-9290

---



∞ The paper used in this book is acid-free and falls within the guidelines
established to ensure permanence and durability.
Visit the AMS home page at URL: http://www.ams.org/

10 9 8 7 6 5 4 3 2 1    05 04 03 02 01 00

# Contents

## Note from the AMS

The three articles in this supplement are part of the book "Methods of Qualitative Theory of Differential Equations and Related Topics" (AMS Translations Series 2, vol. 200, Amer. Math. Soc., Providence, RI, 2000). Due to a technical error, they were not included in the original volume. The AMS apologizes for the resulting inconvenience.

Amer. Math. Soc. Transl.
(2) Vol. **200**, 2000

# Some Topological Structures of Quadratic Systems with at Least Four Limit Cycles

E. A. Andronova

*To the memory of Professor E. A. Leontovich-Andronova*

By a quadratic system we mean a two-dimensional autonomous system of differential equations

$$\dot{x} = P(x,y) = \sum_{i+j=0}^{2} a_{ij}x^i y^j, \qquad \dot{y} = Q(x,y) = \sum_{i+j=0}^{2} b_{ij}x^i y^j, \tag{1}$$

where $P(x,y)$, $Q(x,y)$ are relatively prime real polynomials of degree at most two such that at least one of them is nonlinear. The problem of finding the maximal number and relative position of the limit cycles of system (1) is well known and difficult; it is the second part of Hilbert's 16th problem in the simplest variant of the quadratic case. The conjecture that a quadratic system has at most three limit cycles has held for many years. It was based on the result of Bautin [**1**] who proved that there are quadratic systems with three limit cycles that arise from a weak focus or center (but no more than three). Examples of quadratic systems with four (or more) limit cycles were given in [**2, 3**]. These were sensational results. Investigation of such systems was continued in several papers, for example in [**4, 5, 6, 7**] and others. Note that well-known topological structures with four limit cycles have two foci in the finite part of the plane with distribution of cycles $(3,1)$ and one (saddle) or three (two saddles and one node) singular points at infinity. But recently in the work of Leonov [**8**] two examples of quadratic systems having at least four limit cycles around the origin appeared. The result mentioned above will be considered in the second part of the present paper.

The first part of this work deals with the investigation of quadratic systems with a weak focus of the third order or center. By a change of variables such a system can be reduced to the following form:

$$\dot{x} = -y + kx^2 + 5axy + ny^2, \qquad \dot{y} = x + ax^2 + (3k+5n)xy. \tag{2}$$

The calculation of the three first focus quantities at the origin for system (2) gives the following expressions

$$\bar{v}_3 = \bar{v}_5 = 0, \qquad \bar{v}_7 = \frac{25\pi a}{32}\,[3(k+2n)(k+n)^2 - a^2(5k+6n)](2a^2+2n^2+kn).$$

2000 *Mathematics Subject Classification.* Primary 58F21.

If the quantities

$$(3) \qquad a, \quad 2a^2 + 2n^2 + kn, \quad 3(k+2n)(k+n)^2 - a^2(5k+6n)$$

are all not equal to zero, then system (2) has a weak focus of the third order at the origin. If at least one of the quantities in (3) is equal to zero, then we obtain three different cases in which a center occurs. Note that three is the highest order of a weak focus for quadratic systems on the plane.

System (2) was studied in [9] in which a decomposition of the space of parameters $a$, $k$, $n$ was constructed. Information on the qualitative structures on the bifurcation surfaces $a = 0$, $2a\wedge 2 + 2n\wedge 2 + kn$, $3(k+2n)(k+n)\wedge 2 - a\wedge 2(5k+6n) = 0$ corresponding to the case of a center at the origin was used. A key point was the absence of limit cycles around the focus of multiplicity three [**10, 11**]. The study makes it possible to

1) distinguish domains in the space of parameters $a$, $k$, $n$ for which there exists a system (1) close to system (2), corresponding to some point of such a domain, and possessing at least four limit cycles of type $(3,1)$ with two foci in the finite part of the plane and one singular point (a saddle) or three singular points (two saddle and a node) on the equator of the Poincaré disk;

2) prove the existence of a quadratic system possessing four limit cycles of type $(3,1)$ with two foci in the finite part of the plane and two singular points (saddle-node and saddle) on the equator of the Poincaré disk.

Let us recall some properties of system (2) established in [**9**]. This system has only two foci at the points $O(0,0)$ and $C(0,1/n)$ in the finite part of the plane if the condition $25a^2 + 12n(k+2n) < 0$ holds. To investigate the points at infinity, we make the substitution $z = 1/x$, $v = y/x$ which tranforms system (2) to the form

$$\dot{z} = -kz - 5azv + z^2v - nzv^2, \qquad \dot{v} = a + z + (2k+5n)v - (z-5a)v^2 - nv^3.$$

Singular points at infinity correspond to $z = 0$, and we obtain the equation

$$(4) \qquad a + (2k+5n)v - 5av^2 - nv^3 = 0.$$

Its discriminant has the form

$$D = -125a^4 - (262n^2 + 170kn + 25k^2)a^2 - n(2k+5n)^3.$$

The characteristic equation for the singular points at infinity can be written in the form

$$(k + 5av_i + nv_i^2 + \lambda)((2k+5n) - 10av_i - 3nv_i^2 - \lambda) = 0,$$

where $v_i$ are the roots of equation (4). If $\lambda_1 = -k - 5av - nv^2 = 0$, $\lambda_2 \neq 0$, then condition $3a^2(5k+8n) - k(3k+5n)^2 = 0$ holds. In this case complicated points are obtained as the result of the coincidence of points on the equator on the Poincaré disk with points coming from the finite part of the plane. If $\lambda_1 \neq 0$, $\lambda_2 = (2k+5n) - 10av - 3nv^2 = 0$, then $D = 0$, which corresponds to the merging of singular points on the equator. It is easily seen that expression $(2k+5n) - 10av - 3nv^2$ is the derivative of the right-hand side of (4).

LEMMA. *If the conditions*

1. $D > 0$, $a \neq 0$ *or*
2. $D \leqslant 0$, $a \neq 0$, $-0.4k < n < \bar{n}$, $|a| < -k/12\sqrt{5}$,

*where* $\bar{n} = (-153 \pm \sqrt{6129})k/576$ *and* $k < 0$, *hold, then system* (2) *has at least one limit cycle around the focus at the point* $C(0, 1/n)$.

PROOF. In order to prove this lemma, we use a straight line and a hyperbola transversal to the vector field [**9**]:

$$y = -\frac{1}{3k+5n}, \quad F(x,y) = \frac{n}{2k+5n}\left(y - \frac{2(k+3n)}{n(k+5n)}\right)^2 - x^2 - \frac{2(k+n)(k+3n)}{kn(k+5n)^2} = 0.$$

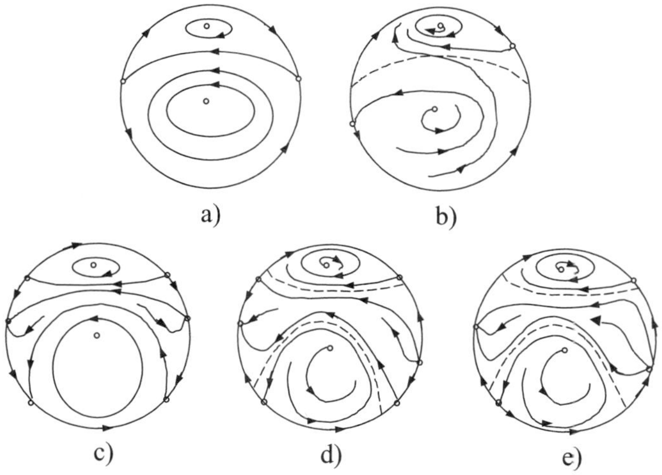


FIGURE 1

Both curves are partial integrals of the center case ($a = 0$) of system (2). The corresponding topological structures are shown in Figures 1a, 1b for the first case of the lemma, and in Figures 1c, 1d, 1e for the second case. Note that according to [**9**], the region where the topological structure explained in Figure 1d exists, is bounded not by the straight line $n = \overline{n}$, but by the dotted line. This line corresponds to the bifurcation surface on which the separatrix cycle surrounding the point $C(0, 1/n)$ and containing the part of the arc of the equator of the Poincaré disk is formed. Conditions which determine this surface are not obtained, but it is shown in [**9**] where the surface is located and how it intersects the surface $D = 0$. On the plane $k = \text{const} < 0$, we have two intersection points

$$M\left(\frac{k}{12\sqrt{5}}, -\frac{5k}{12}\right) \quad \text{and} \quad N\left(-\frac{k}{12\sqrt{5}}, -\frac{5k}{12}\right)$$

(see Figure 2). Moreover, from [**9**] it also follows that the topological structures presented in Figures 1b, 1d, 1e correspond to the case $a > 0$. If $a < 0$, then the topological structures remain as before, but to points on the plane $k = \text{const}$ symmetric with respect to coordinate line $a = 0$ correspond phase portraits symmetric with respect to the $x$-axis, and the direction of motion on the trajectories changes. □

PROPOSITION 1. *If the conditions of the lemma hold, then there exist arbitrarily small changes of the coefficients of system* (2) *such that the perturbed system has four* (*or more*) *limit cycles with distribution* (3, 1) *with two foci in the finite part of the plane and one* (*saddle*) *or three* (*two saddle and one node*) *or two* (*saddle-node and saddle*) *singular points on the equator of the Poincaré disk.*

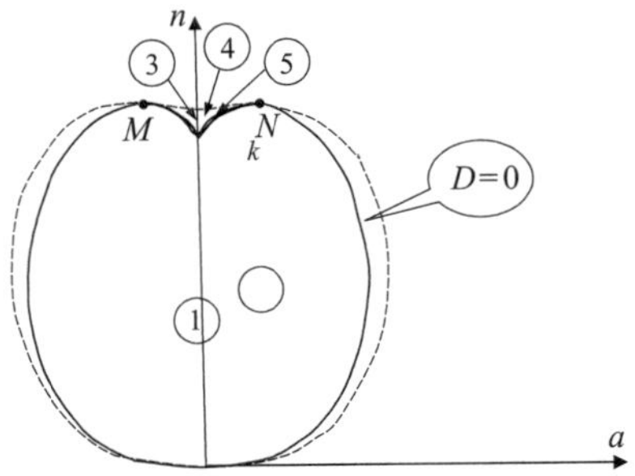


FIGURE 2

PROOF. Let conditions 1, 2 of the lemma hold, but let $D \neq 0$; then the proof is trivial. Since at the point $O(0,0)$ the weak focus is of the third order, we can choose a small perturbation of system (2) so that in a small neighborhood of the origin at least three limit cycles appear in addition to the limit cycle surrounding the point $C(0, 1/n)$. Here the number of singular points on the equator of the Poincaré disk, their character, and the behavior of the separatrix do not change. For example, see the system

$$\dot{x} = \lambda x - y + kx^2 + (5a+\delta)xy + ny^2, \qquad \dot{y} = x + ax^2 + \left(3k + 5n + \frac{\delta(k+n)+\varepsilon}{a}\right)xy$$

from [4], where $0 \ll |\lambda| \ll |\varepsilon| \ll |\delta| \ll 1$, $\delta\varepsilon > 0$, $\varepsilon\lambda > 0$.

Now we consider the remaining case $D = 0$, assuming that the inequalities in condition 2 hold. System (2) has only two singular points on the equator of the Poincaré disk, a saddle and a saddle-node. Since at the origin the weak focus of the third order remains, it follows that as in the case above, we can choose small additions to the coefficients of system (2) so that at least three limit cycles appear in a small neighborhood of the origin. Summing up, there are at least four limit cycles obtained under the perturbation of system (2). Moreover, the equation

$$(2k + 5n) - 10av - 3nv^2 = 0 \tag{5}$$

for the coordinates of the saddle-node on the Poincaré disk changes and, obviously, its roots will change too, as well as the discriminant of equation (4). It is easily seen that the right-hand side of the changed equation (5) is still the derivative of the right-hand side of the changed equation (4). Therefore, in this case the perturbed system has at least four cycles with distribution (3, 1) and saddle-node and saddle on the Poincaré disk.

Let us consider a quadratic system in generic form with the coefficients denoted as in the paper [**8**]:

$$\begin{aligned} \dot{x} &= a_1x^2 + b_1xy + c_1y^2 + \alpha_1x + \beta_1y, \\ \dot{y} &= a_2x^2 + b_2xy + c_2y^2 + \alpha_2x + \beta_2y. \end{aligned} \tag{6}$$

Using the Lienard equation, some interesting results were obtained in [**8**]. The Lienard equation

$$\ddot{x} + f(x)\dot{x} + g(x) = 0 \tag{7}$$

is a powerful tool for studying limit cycles of systems in the plane, in particular, quadratic systems. See the surveys [**12, 13**] and the recent works [**14, 15**].

As usual, the system

$$\dot{x} = y - F(x), \qquad \dot{y} = -g(x)$$

equivalent to equation (7) is considered, where

$$F(x) = \int_0^x f(\xi)\, d\xi.$$

However, in [**8**] a different form was used,

$$\dot{x} = y, \qquad \dot{y} = -H(x)y - G(x), \tag{8}$$

as well as the equivalent equation

$$\frac{dy}{dx} = \frac{-H(x)y - G(x)}{y}. \tag{9}$$

By a change of variables system (6) can be reduced to a form corresponding to system (8), so that instead of studying the limit cycles of system (6) we can study the limit cycles of system (8).

In [**8**] Leonov considered equation (9) for the case in which

$$H(x) = \varepsilon H_0(x), \qquad G(x) = G_0(x) + \varepsilon G_1(x),$$

where $\varepsilon$ is a small parameter. Moreover, if $H(x) = 0$, then system (8) can be integrated and

$$F(x, x_0) = \sqrt{-2\int_{x_0}^x G_0(s)\, ds}$$

is its integral.

Let us present the main theorem from [**8**], on which certain computer experiments were based.

THEOREM (G. A. Leonov). *If the function*

$$\psi(x_0) = \int_{x_0}^{u(x_0)} H_0(x)F(x)\, dx$$

*changes sign $n$ times as $x_0$ ranges from $\tau_1$ to $\tau_2$, then for each sufficiently small $\varepsilon$ system* (8) *has at least $n$ cycles that intersect the set $\{x \in (\tau_1, \tau_2),\ y = 0\}$, where $x_0$, $u(x_0)$ are zeros of the function $F(x)$ such that $F(x) \neq 0$ for all $x \in (x_0, u(x_0))$.*

In [**8**] two examples were presented. In the first example the coefficients were

$$\sigma = 0.00015, \quad \nu_1 = \frac{350}{343}\sigma^2, \quad \nu_2 = -\frac{1625}{98}\sigma^4, \quad c_1 = 0,$$

$$c_2 = \frac{6}{7} - \nu_1, \quad a_1 = \varepsilon\left(1 + \frac{4\cdot 10^{-10}}{1-2c_2}\right), \quad a_2 = -\frac{5}{7}, \quad b_1 = -1,$$

$$b_2 = \varepsilon\left(\frac{3}{1-7\nu_1} + \nu_2 - \frac{8\cdot 10^{-10}}{1-2c_2}\right), \quad \alpha_1 = \frac{14625}{4\cdot 7^5}\sigma^6\varepsilon,$$

$$\alpha_2 = -1, \quad \beta_1 = 1, \quad \beta_2 = -\frac{14625}{4\cdot 7^5}\sigma^6\varepsilon;$$

in the second example,

$$a_1 = -2.52966630809364414\cdot 10^{-6}\varepsilon, \quad a_2 = -2.00001011871642434,$$

$$b_1 = -2, \quad b_2 = 1.00000505933261619\varepsilon, \quad c_1 = 0, \quad c_2 = 2.00000505935821217,$$

$$\alpha_1 = 0, \quad \alpha_2 = -1, \quad \beta_1 = 1, \quad \beta_2 = -1.82135935748424846\cdot 10^{-22}\varepsilon,$$

where $\varepsilon$ is a small parameter. Further, in the above paper, with the help of a computer experiment, it was shown that system (6) has examples of at least four limit cycles surrounding the origin in each of the two examples.

Let us suppose that $\varepsilon = 0$. Then the change $x \mapsto y$ and $y \mapsto x$ transforms system (6) with coefficients from the first and second examples to the form

$$\begin{aligned} \dot{x} &= -y + \left(\frac{6}{7} - \nu_1\right)x^2 - \frac{5}{7}y^2, \\ \dot{y} &= x - xy, \end{aligned} \tag{10}$$

and

$$\begin{aligned} \dot{x} &= -y + 2.00000505935821217x^2 - 2.00001011871642434y^2, \\ \dot{y} &= x - 2xy, \end{aligned} \tag{11}$$

respectively. Both systems (10) and (11) are particular cases of the more general system

$$\dot{x} = -y + kx^2 + ny^2, \qquad \dot{y} = x + bxy \tag{12}$$

with a singular point of center type at the origin. If $b - k \neq 0$, $b - 2k \neq 0$, $k \neq 0$, $b \neq 0$, the integral has the following form:

$$\left[x^2 - \frac{n}{b-k}y^2 + \frac{2(b-k+n)}{(b-k)(b-2k)}y + \frac{b-k+n}{k(b-k)(b-2k)}\right]|(1+by)|^{-2k/b} = C. \tag{13}$$

System (12) was investigated in [**16**] and belongs to the first center case in the classification of [**16**]. There system (12), its bifurcation surfaces, phase portraits on the Poincaré disk, and their location in the parameter space were considered. The bifurcational diagram is presented for $b = \text{const} < 0$ and $b = 0$. See Figure 3 for $b = \text{const} < 0$, where $n = 0$, $k = 0$, $b + n = 0$, $b - k = 0$ and $k + n = 0$ are the bifurcational lines. The numbers in the small cycles correspond to phase portraits shown in Figures 1a, 1c, 4a, 4b, 4c, and Figures 4a, 4c correspond to systems (10) and (11) respectively. Hence the systems obtained from (10) and (11) under perturbations as in the Leonov examples possess topological structures with at least four limit cycles surrounding the origin, having three saddles in the finite part of the plane and three nodes on the equator of the Poincaré disk, but with different behavior of the separatrix of the saddles. It is easy to visualize topological structures in Figure 4a and Figure 4c, taking into account the fact that the foci at the origin are stable and $y = 1/2$ is a transversal line for the trajectory of the second example.

It is expected that such systems with at least four limit cycles may be obtained similarly from system (12) by perturbation if $(n + k)n < 0$ or $(n + k)n < 0$ and $k \cdot n < 0$. But the conjecture mentioned above is not supported by rigorous mathematical considerations; it is based only on Leonov's computer experiments, who has written the following in his paper: "It must be emphasized that these computer experiments require a mathematically rigorous proof". □

It is difficult to obtain a rigorous proof and especially to define bifurcations which yield limit cycles. Where do the limit cycles appear from in the situation considered? They can appear from a singular point, from a separatrix contour, or from an integral curve surrounding a singular point of center type.

It is easy to see that the center at the origin of system (12) has cyclicity equal to two if $k+n \neq 0$ and $b-3k-5n \neq 0$. Moreover, according to [**17**], the cyclicity of any

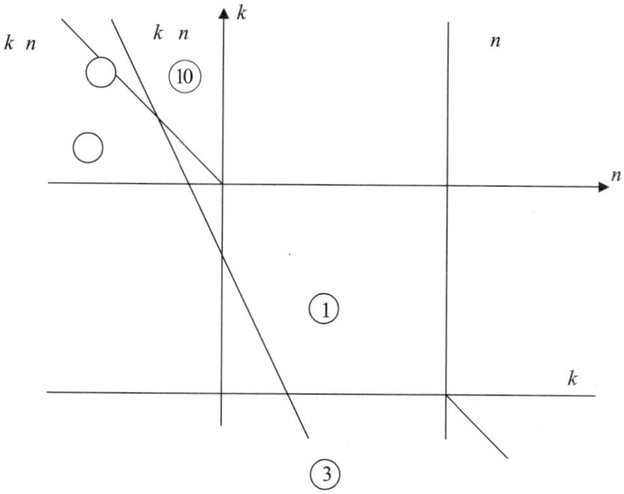


FIGURE 3

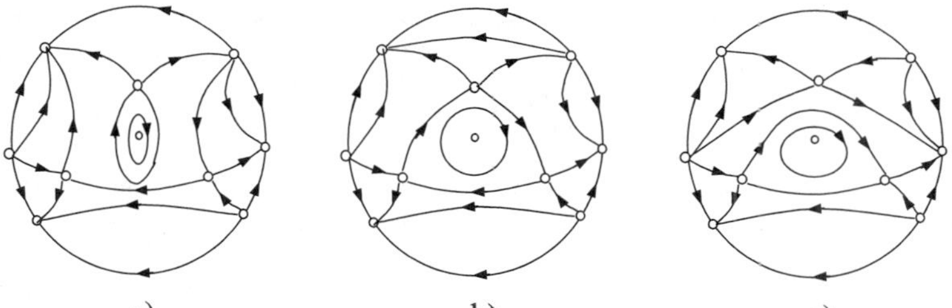


FIGURE 4

bounded segment is also equal to two. If $k + n = 0$ and $k \cdot n < 0$, then system (12) has a phase portrait topologically equivalent to that shown in Figure 4b. Note that in this case the cyclicity of the center becomes equal to one, if $b - 3k + n \neq 0$, but the cyclicity of the separatrix triangle is equal to three [**17**].

In our opinion a very interesting situation occurs when $k + n = 0$ and $b - 3k - 5n = 0$ (see the point A in Figure 3). In this case three center conditions hold simultaneously (see [**16, 6**]) and the cyclicity of the center of system (12) is equal to three, while the cyclicity of the separatrix triangle also remains equal to three. In addition, the integral (13) becomes analytical.

The author express his thanks A. D. Bruno and M. W. Dolov for their interest in the topic and for helpful discussions.

## References

[1] N. N. Bautin, *On the number of limit cycles which appear with the variation of the coefficients from an equilibrium position of focus or center type*, Math. Sb. **30** (1952), 181–196; English transl., Amer. Math. Soc. Transl. Ser. 2, vol. 15, Amer. Math. Soc., Providence, RI, 1962, pp. 396–414.

[2] L. Chen and M. Wang, *The relative position and number of limit cycles of the quadratic differential systems*, Acta Math. Sinica **22** (1979), no. 6, 751–758.

[3] S. Shi, *A concrete example of the existence of four limit cycles for the plane quadratic systems*, Sci. China Ser. A **23** (1980), no. 2, 153–158.
[4] ———, *A method of constructing cycles without contact around a weak focus*, J. Differential Equations **41** (1981), no. 3, 301–312.
[5] Y. Qin, S. Shi, and B. Cai, *On limit cycles of planar quadratic systems*, Sci. China Ser. A **26** (1982), no. 1, 41–50.
[6] E. A. Andronova, *On the topology of quadratic systems with four (or more) limit cycles*, Uspekhi Mat. Nauk **41** (1986), no. 2, 183–184; English transl., Russian Math. Surveys. **41** (1986), no. 2, 191–192.
[7] D. W. Litvinenko, *Quadratic systems with four limit cycles and with three singular points at the infinity*, 7th Czechoslovac Conf. on Diff. Equations and Appl., Abstracts I, Praha, 1989, p. 127.
[8] G. A. Leonov, *Two-dimensional quadratic systems as a Lienard equation*, International Symposium on Qualitative Theory of Diff. Equat., preprint of the lecture, Holland, Delft, 1995.
[9] E. A. Andronova, *A partition of the parameter space of a quadratic system with a triple focus or a center at the origin*, Differentsial′nye Uravneniya **34** (1998), no. 4, 441–450; English transl., Differential Equations. **34** (1998), no. 4, 437–446.
[10] L. A. Cherkas, *The absence of the limit cycles around the weak focus of the third order for the quadratic systems on the plane*, Differentsial′nye Uravneniya **22** (1986), no. 11, 2015–2017; English transl. in Differential Equations, **22** (1986).
[11] C. Li, *Nonexistence of a limit cycle around a week focus of order three for any quadratic system*, Chinese Ann. Math. Ser. B **7** (1986), 174–190.
[12] L. A. Cherkas, *Methods of estimating the number of limit cycles of autonomous systems*, Differentsial′nye Uravneniya **13** (1977), 779–802; English transl. in Differential Equations **13** (1977).
[13] U. Staude, *Uniqueness of periodic solutions of the Lienard equations*, Recent Adv. Differential Equations, Academic Press, New York, 1981, pp. 421–429.
[14] W. A. Coppel, *A new class of quadratic systems*, J. Differential Equations **2** (1992), 360–372.
[15] Xianwu Zeng, Zhang-Zhi-Fen, and Suzhi Gao, *On the uniqueness of the limit cycle of the generalized Lienard equation*, Bull. London Math. Soc. **26** (1994), 2–47.
[16] E. A. Andronova, *The distribution of the parameter space of the quadratic system with the singular point of the center type*, Gor′kiĭ, VINITI, no. 1093–B87, pp. 2–47. (Russian)
[17] H. Zoladek, *The cyclicity of triangles and segments in quadratic systems*, J. Differential Equations **122** (1995), no. 1, 137–159.

MATHEMATICS DEPARTMENT, VOLGA STATE ACADEMY OF WATER TRANSPORTATION, 5 NESTEROVA STREET, 603600, NIZHNY NOVGOROD, RUSSIA

Translated by THE AUTHOR

Amer. Math. Soc. Transl.
(2) Vol. **200**, 2000

# Torus Maps from Weak Coupling of Strong Resonances

John Guckenheimer and I. Khibnik

ABSTRACT. This paper investigates a family of diffeomorphisms of the two-dimensional torus derived from a model of an array of driven, coupled lasers (see Braiman et al. [**2**]). It gives a negative answer to a question raised by Baesens et al. (see [**3**]) about two-parameter families of diffeomorphisms and flows of the two-dimensional torus by exhibiting resonance regions in which the two parameter family has no Hopf or Takens–Bogdanov points. The example can be interpreted as a general model for single-mode coupling of two strongly nonlinear oscillators driven near resonance. We give a complete analysis of the steady state bifurcations in this model that depends on four parameters and a description of the observed global bifurcations in a two-parameter subfamily.

## 1. Introduction

The phase dynamics of a system of $n$ oscillators take place on the $n$-dimensional torus. Coupled oscillators exhibit *mode locking* and *resonance* in which trajectories approach a limit that lies on a torus whose dimension is smaller than $n$. For a pair of oscillators ($n = 2$), the geometry associated with mode-locking is quite well understood. Flows on the two-dimensional torus with global cross-sections have return maps that are diffeomorphisms of a circle. The theory of rotation numbers characterizes the flows that exhibit mode-locking as those whose return maps have rational rotation numbers. Rotation numbers are defined as elements of $\mathbb{R}/\mathbb{Z}$, or equivalently can be regarded as real numbers whose integer part is determined by the lift of a diffeomorphism of the circle to a diffeomorphism of the line (Herman [**6**]). The work of Arnold [**1**], Herman [**6**], and Yoccoz [**8**] gives a very thorough analysis of the dynamics of families of diffeomorphisms of the circle. We shall say that a pair of oscillators has a *strong resonance* if the corresponding circle diffeomorphism has a fixed point, or equivalently the oscillators are mode-locked with a rotation number that is an integer. There is a close correspondence between the analysis of diffeomorphisms of the circle with fixed points and the theory of flows of the circle. For diffeomorphisms that are close to rotations, this correspondence has a simple expression in terms of the Euler integration algorithm: the time $\varepsilon$ map of the equation $\dot{x} = f(x)$ is approximated by the diffeomorphism $F(x) = x + \varepsilon f(x)$. The equilibrium points of $f$ and the fixed points of $F$ are identical. Using a similar

2000 *Mathematics Subject Classification.* 58F14.

correspondence, we can learn much about mode-locking of several oscillators by approximating their return maps by flows.

The mathematical theory of families of torus flows and diffeomorphisms in higher dimensions is less complete. Baesens et al. [**4**] defined resonance regions of a family of torus diffeomorphisms to be the set of parameters for which the system has periodic orbits. They studied the bifurcations within and on the boundary of resonance regions of two parameter families of two-dimensional torus maps. Their starting point was the family of *twist maps*, a model for uncoupled oscillators in which the relative frequencies of the oscillators are parameters. They explored nonlinear perturbations of twist map families as models of the typical behavior of weakly coupled oscillators. They gave extensive descriptions of the topology of resonance regions in the parameter spaces and of the global bifurcations that interacted with the resonance regions. The resulting picture displays the structure of a higher dimensional analog of a resonance tongue for a family of circle diffeomorphisms. The dynamics of diffeomorphisms of the two-dimensional torus is much less constrained than that of the circle, so there is much more variability in the bifurcation diagrams of the resonance regions. Figure 1 shows the bifurcation diagram of a two parameter family of flows on the two torus that corresponds to a *simple resonance region*. A simple resonance region of a family of diffeomorphisms is one in which no parameter value yields more than two periodic orbits of the resonance type of the region. Figure 1 displays several global bifurcation curves at which the number of periodic orbits of the flows change. The diagram is *generic* in the sense that all of the singular points of the bifurcation diagram are codimension two bifurcations of flows whose qualitative structure is unchanged by perturbations of the family of flows. (Perturbations of the time $t$ map of these flows as a family of diffeomorphisms introduce much additional complexity, including the presence of chaotic invariant sets.) In all of the cases they examined, the resonance regions of torus maps contained codimension one Hopf bifurcations and codimension two *Takens–Bogdanov* bifurcations of periodic orbits. The Takens–Bogdanov bifurcations of diffeomorphisms are characterized as periodic orbits for which the linearization has one as an eigenvalue of algebraic multiplicity two. Baesens et al. [**3**] proved that Takens–Bogdanov points must occur in the boundaries of simple resonance regions and asked the question as to whether all resonance regions have Takens–Bogdanov points.

This paper analyzes a family of flows of the two-dimensional torus without Hopf bifurcations or Takens–Bogdanov bifurcations. While our results are about flows, they translate directly into comparable results about diffeomorphisms through the correspondence between flows and the Euler time $\varepsilon$ step of the flows. Thus, we give a negative answer to the question asked above. More specifically, if $\dot{x} = f(x)$ defines a flow on the torus $T^2$, then the equilibrium points of $f$ are the fixed points of the diffeomorphisms $F(x) = x + \varepsilon f(x)$ for all $0 < \varepsilon < \sup(|Df(x)|^{-1})$. Moreover, the stability of the equilibrium points of $f$ and the fixed points of $F$ agree. The examples we analyze arose in the study of solid-state laser arrays where entrainment is due to external periodic forcing (Braiman et al. [**2**], Khibnik et al. [**7**]. They are defined by the equations

$$\begin{aligned} \dot{x_1} &= a + b - c\sin x_1 - d\sin(x_2 - x_1), \\ \dot{x_2} &= a - b - c\sin x_2 - d\sin(x_1 - x_2), \end{aligned} \tag{1}$$

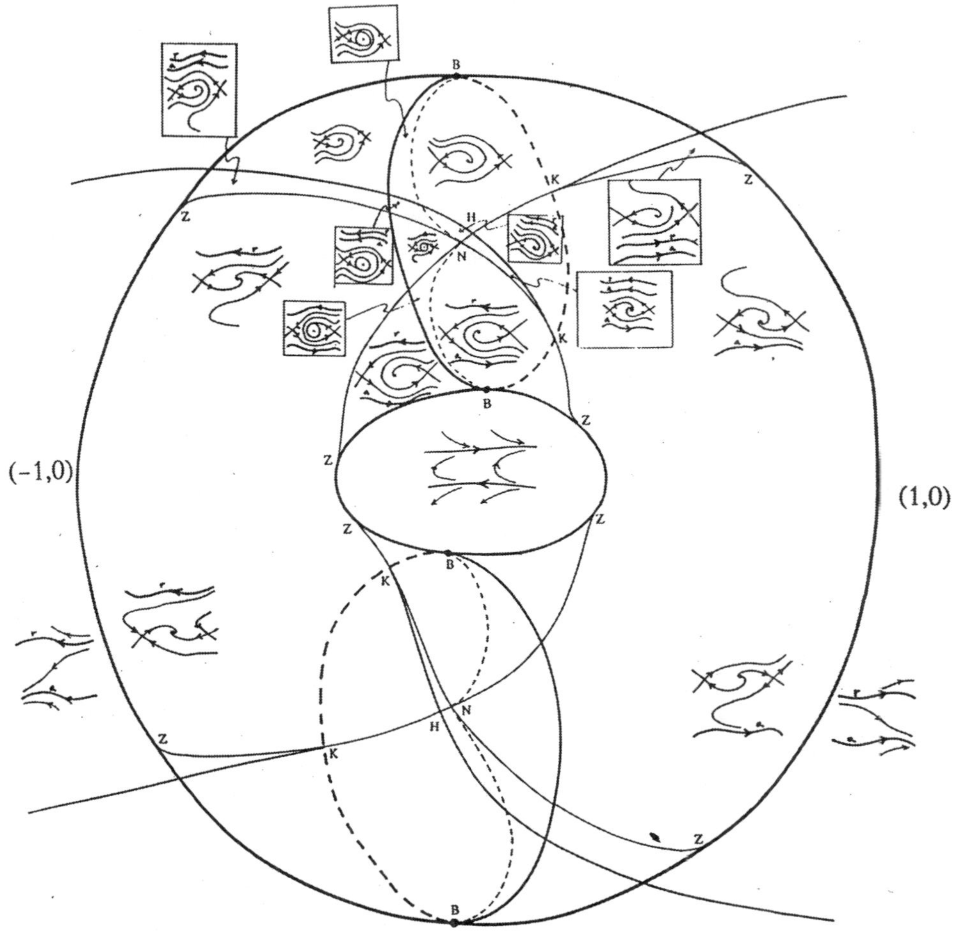


FIGURE 1. This figure, reproduced from Baesens et al. [**4**], shows the geometry of a simple resonance region of a two parameter family of torus flows. In the interior of an annulus in the parameter plane, there are two equilibrium points of the flows. Phase portraits for parameters in different regions are superimposed upon the bifurcation diagram.

where we fix $c$ and $d$ and regard $a$ and $b$ as the active parameters of the family. In the laser models, $a$ and $b$ represent the frequencies of the two lasers, $c$ is the amplitude of external forcing and $d$ is the coupling strength. These equations are derived by applying singular perturbation theory to a more complex model and then restricting the reduced equations to an invariant torus. The single equation $\dot{x} = a + c \sin x$ is the result of applying this procedure to a single resonantly forced oscillator. This is the vector field analog of a family of diffeomorphisms displaying strong resonance. When $|a| \leqslant |c|$, the system has equilibrium point(s) corresponding to strongly resonant oscillations. When $|a| > |c|$, there are no equilibria and the system is no longer resonant. The system we study represents two such oscillators with slightly different resonant frequencies. The frequency difference is represented by the parameter $b$. The two oscillators are coupled by a term that depends symmetrically upon the phase difference of the two oscillators. We have not tried to pinpoint other physical problems that lead to this system, but we think that its behavior is typical of symmetry breaking perturbations of weakly coupled, resonantly forced oscillations of two identical oscillators.

Rescaling time in the vector fields (1) multiplies each of the parameters by a scalar. Thus the bifurcation set of the family has a conical structure, and it suffices to study a three-parameter family to determine all of the dynamics of the system. For example, setting $c^2 + d^2 = 1$ is a normalization that fixes the size of the resonance region in the $(a, b)$ plane and allows us to easily explore both the limits of strong coupling of weakly forced oscillators ($c = 0$) and weak coupling of strongly forced oscillators ($d = 0$). Here we focus upon global bifurcations that occur for weak coupling of strongly forced oscillators ($|d/c| \ll 1$.)

## 2. Local bifurcations

This section discusses the local bifurcations of the two parameter family of vector fields (1) obtained by varying $a$ and $b$. For each value of $c$ and $d$, the local bifurcations form curves in the $(a, b)$ parameter space that we describe. There are three types of local codimension one bifurcations of equilibria: saddle-nodes, Hopf bifurcations and resonant saddles. Resonant saddles are not always regarded as bifurcations since they do not affect the local topological properties of a phase portrait near an equilibrium point, but they play an important role in some kinds of global bifurcations. Since they are significant for the global bifurcations of the family (1), we include their analysis here. The different local bifurcations occur when the Jacobian matrix $J$ of equations (1)

$$\begin{bmatrix} -c\cos x_1 + d\cos(x_1 - x_2) & -d\cos(x_1 - x_2) \\ -d\cos(x_1 - x_2) & -c\cos x_2 + d\cos(x_1 - x_2) \end{bmatrix}$$

has a zero eigenvalue, pure imaginary eigenvalues or a pair of real eigenvalues whose sum is zero, respectively. Zero eigenvalues occur when

$$\det(J) = c^2 \cos x_1 \cos x_2 - cd\cos(x_1 - x_2)(\cos x_1 + \cos x_2) = 0$$

and pure imaginary or resonant eigenvalues imply that

$$\operatorname{trace}(J) = 2d\cos(x_1 - x_2) - c(\cos x_1 + \cos x_2) = 0.$$

Pure imaginary and real, resonant eigenvalues are distinguished by the sign of $\det(J)$: imaginary eigenvalues occur when $\det(J) > 0$ and real, resonant eigenvalues occur when $\det(J) < 0$. When $\det(J)$ and $\operatorname{trace}(J)$ are both zero, we say that the *Takens–Bogdanov bifurcation* occurs. This is a codimension two bifurcation. Our first result has a simple proof, but it answers negatively a question posed by Baesens et al. [**3**].

THEOREM 1. *Assume that $c \neq 0$ and $d \neq 0$. The vector field defined by*

$$\dot{x_1} = a + b - c\sin x_1 - d\sin(x_2 - x_1),$$
$$\dot{x_2} = a - b - c\sin x_2 - d\sin(x_1 - x_2)$$

*has no Hopf or Takens–Bogdanov bifurcations as $a$ and $b$ vary.*

PROOF. Assume that $c$ and $d$ are both nonzero. The curve defined by $\operatorname{trace}(J) = 0$ consists of points with either pure imaginary eigenvalues (Hopf points) or points at which there are real eigenvalues of opposite sign and equal magnitude (resonant saddles). When trace$(J) = 0$, $J$ is readily seen to have negative determinant since

$$\cos(x_1 - x_2) = \frac{c(\cos x_1 + \cos x_2)}{2d}$$

implies

$$\begin{aligned} \det(J) &= c^2\left(\cos x_1 \cos x_2 - \frac{(\cos x_1 + \cos x_2)^2}{2}\right) \\ &= -\frac{c^2}{2}((\cos x_1)^2 + (\cos x_2)^2) < 0 \end{aligned}$$

unless $\cos x_1 = \cos x_2 = 0$. However, $\cos x_1 = \cos x_2 = 0$ implies $|\cos(x_1 - x_2)| = 1$, contradicting the equation trace$(J) = 0$. Thus trace$(J) = 0$ implies that $\det(J) < 0$ and that the only points where trace$(J) = 0$ are resonant saddles. This proves the theorem. □

We next examine the geometry of the bifurcation curves in more detail, beginning with the resonant saddles. The gradient of trace$(J)$ is

$$(c \sin x_1 - 2d \sin(x_1 - x_2), c \sin x_2 + 2d \sin(x_1 - x_2)).$$

Points where $c \neq 0$ and the gradient vanishes satisfy $\sin x_1 + \sin x_2 = 0$. This implies that $x_2 = x_1 + \pi$ or $x_2 = -x_1$. In the first case, we have $\cos x_1 + \cos x_2 = 0$ and trace$(J) = 0$ implies that $d = 0$. In the second case,

$$\begin{aligned} \cos x_2 = \cos x_1, \quad \cos(x_1 - x_2) &= 2\cos^2 x_1 - 1, \\ \sin(x_1 - x_2) &= 2 \cos x_1 \sin x_1. \end{aligned}$$

Substitution of these expressions into the equations

$$\begin{aligned} \text{trace}(J) = 2d\cos(x_1 - x_2) - c(\cos x_1 + \cos x_2) &= 0, \\ c \sin x_1 - 2d \sin(x_1 - x_2) &= 0 \end{aligned}$$

yields $-2c\cos x_1 + 4d(\cos x_1)^2 - 2d = 0$ and $\sin x_1(c - 4d\cos x_1) = 0$. If $cd \sin x_1 \neq 0$, this pair of equations has no solution. We conclude that the singularities of the curves defined by trace$(J) = 0$ occur when $x_1 = x_2 = 0$ and $c = d$ or when $x_1 = x_2 = \pi$ and $c = -d$. A simple computation of the Hessian of trace$(J)$ shows that the singular points are index 1 Morse critical points of the function trace$(J)$. For $|c| > |d|$, we assert that the set of resonant saddles is a connected curve of trivial homotopy type on the torus of the $(x_1, x_2)$ variables. To see this, note that, when $d = 0$, the solutions of trace$(J) =$ are given by $\cos x_1 + \cos x_2 = 0$ or $x_1 = \pi \pm x_2$. Singularities of this curve occur at $(\pi, 0)$ and $(0, \pi)$. When $d > 0$ is small, the deformations of the solutions near the singularities are readily computed by expanding the cosine functions as Taylor series to obtain the approximate equations $4d = x_1^2 - (x_2 - \pi)^2$ near the point $(0, \pi)$ and $4d = -(x_1 - \pi)^2 + x_2^2$ near the point $(\pi, 0)$. If $|c| < |d|$, the set of resonant saddles has two components, each of which is a closed curve of homotopy type $(1, 1)$ on the $(x_1, x_2)$ torus. This is readily seen by setting $c = 0$ in the equation trace$(J) = 0$ to obtain the curves given by $x_1 - x_2 = \pm\pi$. The following theorem summarizes this discussion.

THEOREM 2. *Assume $|c| > |d|$. As $(a, b)$ vary, the curve of resonant saddle points of vector fields (1) is connected and nonsingular. The curve of resonant saddle points has two nonsingular components if $|c| < |d|$. If $|c| = |d|$, the curve of resonant saddle points has a double point.*

The geometry of the saddle-node bifurcation curves is somewhat more complex than that of the resonant saddle curves. We prove the following result.

THEOREM 3. *For $|c/d| > 2$, the saddle-node curve of the family (1) has two components. For $0 < |c/d| < 2$, the saddle-node curve has three components. All the components are homotopically trivial and have nonsingular projections onto the $(x_1, x_2)$ torus.*

PROOF. The defining equation for saddle-node bifurcations is

$$\det(J) = c^2 \cos x_1 \cos x_2 - cd \cos(x_1 - x_2)(\cos x_1 + \cos x_2) = 0.$$

The gradient of $\det(J)$ is

$$\begin{aligned}(&-c^2 \sin x_1 \cos x_2 + cd(\sin(x_1 - x_2)(\cos x_1 + \cos x_2) + \cos(x_1 - x_2)\sin x_1),\\ &-c^2 \sin x_2 \cos x_1 - cd(\sin(x_1 - x_2)(\cos x_1 + \cos x_2) + \cos(x_1 - x_2)\sin x_2)).\end{aligned}$$

Thus, the singularities of the saddle-node bifurcation curves in their domains occur when the $3 \times 2$ matrix

$$\begin{pmatrix} \cos x_1 \cos x_2 & \cos(x_1 - x_2)(\cos x_1 + \cos x_2) \\ -\sin x_1 \cos x_2 & \sin(x_1 - x_2)(\cos x_1 + \cos x_2) + \cos(x_1 - x_2)\sin x_1 \\ -\sin x_2 \cos x_1 & \sin(x_2 - x_1)(\cos x_1 + \cos x_2) - \cos(x_1 - x_2)\sin x_2 \end{pmatrix}$$

has rank one. Computing the minors of this matrix and using $\sin^2 x_i = 1 - \cos^2 x_i$ yields the following pair of equations:

$$(2)\qquad \begin{aligned}&\cos^2 x_1 \cos x_2 \sin x_1 - \cos^3 x_1 \sin x_2 - \cos x_2 \sin x_2 = 0,\\ &\cos^2 x_2 \cos x_1 \sin x_2 - \cos^3 x_2 \sin x_1 - \cos x_1 \sin x_1 = 0.\end{aligned}$$

If $\sin x_1$ and $\sin x_2$ are not both zero, then eliminating $\sin x_1$ and $\sin x_2$ from equations (2) yields

$$(3)\qquad \cos^4 x_1 + \cos x_1 \cos x_2 + \cos^4 x_2 = 0.$$

Using $\sin^2 x_i = 1 - \cos^2 x_i$ once again to eliminate $\sin x_i$ from the first equation of the system (2), we obtain

$$(4)\qquad \begin{aligned}&\cos^4 x_1 \cos^2(x_2) + 2\cos^3 x_1 \cos^3 x_2 + \cos^4 x_1\\ &\quad = \cos^6 x_1 + \cos^2 x_1 + 2\cos^3 x_1 \cos x_2.\end{aligned}$$

Using the computer program Maple to simultaneously solve equation (3) and equation (4) produces three solutions:

$$\cos x_1 = \cos x_2 = 0, \qquad \cos x_1 = -\cos x_2 = \pm\sqrt{1/2}$$

and solutions with $\cos x_1$ a root of the polynomial $p(z) = z^8 - 3z^6 + 4z^4 - 2z^2 + 1$. All roots of $p(z)$ are complex, so there are no roots in the third family. If $\cos x_1 = \cos x_2 = 0$, then $\cos(x_1 - x_2) = \pm 1$ and hence $d = 0$. If $\cos x_1 = -\cos x_2 = \pm\sqrt{1/2}$, then $\cos(x_1 - x_2) = 0$ and hence $c = 0$. We conclude that the only singularities of the saddle-node curves occur when $\sin x_1$ and $\sin x_2$ are both zero. In this case $\det(J) = 0$ implies that $c/d = \pm 2$. Calculating the Hessian of $\det(J)$, we find that this is a Morse singularity of index one for the zero level set of the function $\det(J)$.

We finish the proof of the theorem by perturbation calculations from the families with $c = 0$ and $d = 0$ that determine the number of components of the saddle node curves in the $(x_1, x_2)$ torus. For $d = 0$ and $c > 0$, the 0 level curve of $\det(J) = 0$ consists of the four circles $x_i = \pm\pi/2$. An easy calculation shows that the function $\det(J)$ is a Morse function in the neighborhood of each of the intersection points $(x_1, x_2) = (\pm\pi/2, \pm\pi/2)$. We proved that the saddle-node curves for $0 < d \ll c$ are nonsingular, so they will be approximated by hyperbolas near $(x_1, x_2) = (\pm\pi/2, \pm\pi/2)$ when $d > 0$ is small. To determine the topology of the saddle-node curves, we only need to determine which direction the crossing curves for $d = 0$ split as we perturb to $c > 0$. Expanding the Taylor series of $\det(J)$ near the points $(x_1, x_2) = (\pm\pi/2, \pm\pi/2)$, we find that the saddle-node curve becomes two homotopically trivial closed curves, one surrounding the origin and one surrounding the point $(x_1, x_2) = (\pi, \pi)$. The perturbation arguments for the case $0 < c \ll d$ are similar. When $c = 0$, the 0 level curve of $\det(J) = 0$ is given by four circles $x_1 \pm x_2 = \pi$ and $x_1 - x_2 = \pm\pi/2$. There are six points of intersection of these circles at $(x_1, x_2) = (\pi/4, 3\pi/4), (3\pi/4, \pi/4), (-\pi/4, -3\pi/4), (-3\pi/4, -\pi/4), (\pi, 0)$ and $(0, \pi)$. The function $\det(J)$ is a Morse function at each of these points. As $c$ is perturbed to be positive, the saddle-node curve becomes three homotopically trivial closed curves which contain the points $(11\pi/8, 3\pi/8), (3\pi/8, 11\pi/8)$ and $(0, 0)$ in their interiors.

Since the saddle-node curves lie at singularities of the map from the $(x_1, x_2)$ torus to the $(a, b)$ parameter plane, additional singularities appear when the curves are projected into the $(a, b)$ parameter plane. The singularity in the $(x_1, x_2)$ torus that occurs when $c/d = \pm 2$ is mapped to a curve that has the local structure of the level set of $u^2 = v^4$ in suitable coordinates. Furthermore, cusps and swallowtails appear along the saddle-node curves in the $(a, b)$ plane. These do not seem to be simple to compute exactly, so we report observations from numerical calculations of the saddle-node curves. The outer component of the saddle-node curve in the $(a, b)$ plane appears to be convex and nonsingular for all values of $c \neq 0$ and $d \neq 0$. For $|c/d| > 3$, the inner component of the saddle-node curve has four cusps. These are located on the $a$ and $b$ coordinate axes. When $|c/d| = 2\sqrt{2}$, there are a pair of swallowtail singularities that occur on the $a$ axis. For $|c/d| < 2\sqrt{2}$, the inner component of the saddle node curve has eight cusps, of which two lie on the $b$ axis and two lie on the $a$ axis. As $|c/d|$ decreases through 2, the two cusps on the $a$ axis merge with one another and branches of the two cusps reconnect smoothly with one another. For $0 < |c/d| < 2$, there are two inner components of the saddle-node curve that are reflections of one another in the $a$ axis. Each of these inner components has three cusps. Additionally, the two components intersect. Denote

$$c_t = \sqrt{2}(\sqrt{5} - 2)\sqrt{\sqrt{5} + 1} \approx 0.60057.$$

There are six points of intersection between the inner two components if $c_t < |c/d| < 2$. If $0 < |c/d| < c_t$, there are only two points of intersection between these curves. This completes our description of the local bifurcations of the equilibria for this family of vector fields. Figure 2 is a collection of diagrams showing the saddle-node and resonant saddle curves in the $(a, b)$ plane for different values of $c/d$. □

## 3. Periodic orbits and global bifurcations

We begin the discussion of periodic orbits of the family (1) by discussing what happens for large values of $(a, b)$. The return maps of torus flows with global cross-sections (and hence no equilibrium points) have rotation numbers (see Guckenheimer and Holmes [**5**]). The rotation number of a return map determines whether

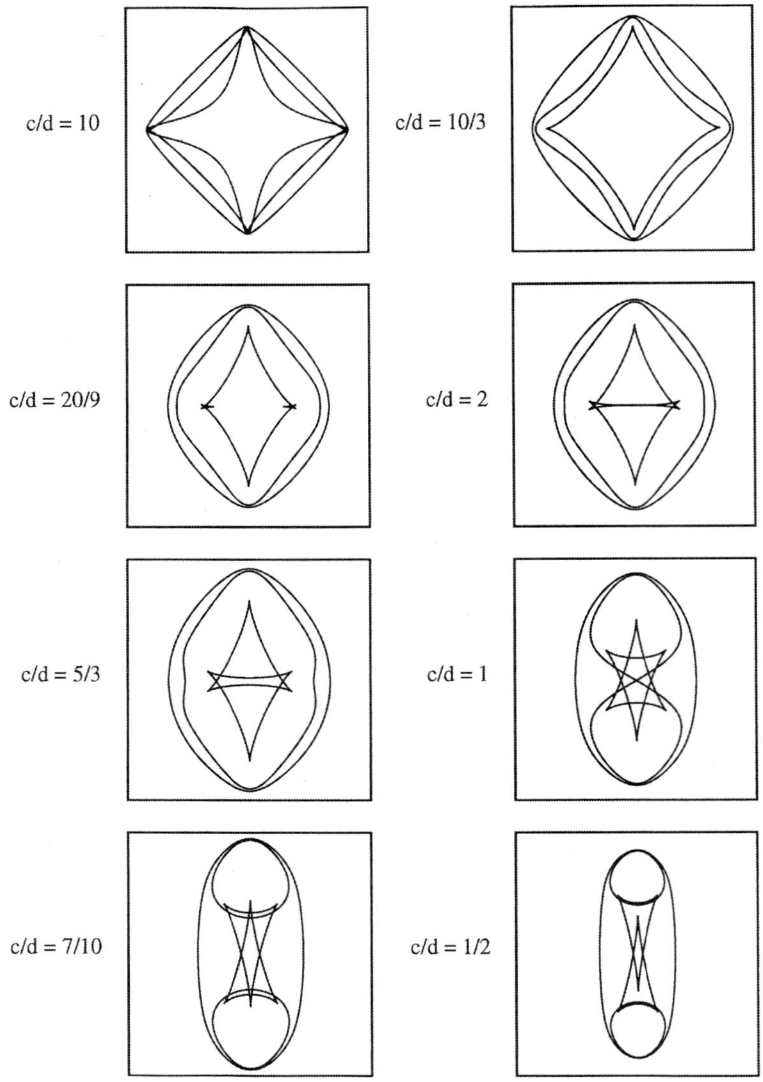


FIGURE 2. Bifurcation diagrams of saddle-node and resonant saddle bifurcations of the families with varying $a$ and $b$ are displayed. The ratio $c/d$ appears to the left of each diagram.

the flow has periodic orbits and characterizes the homotopy types of those orbits. More precisely, while traversing a large circle $a^2 + b^2 = r^2$ in the $(a, b)$ plane, the flows are perturbations of rotations whose return maps assume all rotation numbers. Heuristically, we picture a family of "Arnold tongues," one for each rational number, that extend into the $(a, b)$ plane from infinity. Formally, we define the $(p, q)$ *Arnold tongue* to be the set of flows that have a periodic orbit of homotopy type $(p, q)$. The tongues are disjoint from one another. The primary objective of this section is to describe the boundaries of the Arnold tongues. Our results are not rigorous, but rather an interpretation of numerical computations based upon the assumption that the two parameter family (1) with varying $a$ and $b$ has only generic bifurcations for most values of $c$ and $d$. Here we focus on values of $c/d > 3$ for which the saddle-node curve has only two components and four cusp points and report the results of these investigations. A more complete investigation is required to characterize other cases and the codimension three bifurcations that occur with smaller values of $c/d$.

We discuss briefly the case $d = 0$, viewing this as a singular limit as $c/d$ increases. When $d = 0$, the system (1) separates into two uncoupled equations. The resonance region is the square $|a| + |b| = 1$. All points on the boundary of the resonance region except its corners are flows that have one saddle node equilibrium point. In the interior of the resonance region, all flows are "fully mode locked" with four equilibrium points and no periodic orbit. Outside the resonance region in the strip $|a+b| \leqslant 1$, the flow has periodic orbits of homotopy type $(0, \pm 1)$. Outside the resonance region in the strip $|a - b| \leqslant 1$, the flow has periodic orbits of homotopy type $(\pm 1, 0)$. When $|a + b| > 1$ and $|a - b| > 1$, the flow has a rotation number $\rho \neq 0$. The corners of the saddle-node curve are in the closure of all the Arnold tongues in the quadrant of $|a \pm b| > 1$ adjacent to the corner.

Now assume that $c/d > 3$. The local bifurcation diagram of the $(a, b)$ plane is shown in the top row of Figure 2. The resonance region is bounded by a nonsingular curve of saddle node bifurcations $S_o$ that lies close to the square $|a| + |b| = 1$. There is a second saddle-node curve $S_i$ inside $S_o$ with four cusps located on the coordinate axes. The curve $R$ of resonant saddles lies between the two saddle-node curves near the cusp points, though much of $R$ lies inside $S_i$ if $c/d$ is large enough. The Arnold tongues are perturbations of the Arnold tongues for $d = 0$. Apart from the tongues with homotopy types $(\pm 1, 0)$ and $(0, \pm 1)$, they all collect in the vicinity of the cusps. To make our discussion more concrete, we focus on the region near the cusp on the positive $b$ axis in the $(a, b)$ plane. The structure of the bifurcations near the other cusps is similar. The largest tongues near this cusp have homotopy types $(1, 0)$, $(0, -1)$ and $(1, -1)$. The tongues with homotopy types $(1, 0)$ and $(0, -1)$ are wide and separate the tongues with homotopy types $p > 0 > q$ from those that approach the other cusps. The $(1, -1)$ tongue is special because it is the only tongue that intersects the interior of $S_i$ near the positive $b$ axis.

The bifurcation diagram is symmetric with respect to the $b$ axis, so we further restrict our attention to the left side of the $b$ axis where the tongues have homotopy types with $p < -q$. Arnold tongues $A_{(p,q)}$ with homotopy type $(n, -(n + 1))$ have a somewhat different structure from those with homotopy types $(p, q)$ with $-q > p + 1$. Figure 3 portrays the structure of these tongues. Our first comments apply to all of the Arnold tongues $A_{(p,q)}$ without restrictions on $(p, q)$. The boundary of $A_{(p,q)}$ is formed from curves of codimension one bifurcations with endpoints at codimension two bifurcations. There are two types of codimension one bifurcations that terminate families of periodic orbits: saddle-nodes of periodic orbits and homoclinic bifurcations. Four types of codimension two bifurcations appear on the boundary of these tongues: saddle-node loops, resonant homoclinic bifurcations, transversal intersections of saddle-node bifurcations of equilibria with saddle-node bifurcations of periodic orbits and saddle-node fan points. The intersection of each tongue $A_{(p,q)}$ with $S_o$ is a connected arc $S_{o(p,q)}$ bounded by two codimension two points at which there are transverse intersections of $S_o$ with curves of saddle-nodes of periodic orbits that bound $A_{(p,q)}$ in the exterior of $S_o$. In addition, there are a pair of saddle-node loop points $s_h$ in $S_{o(p,q)}$ at which the saddle-node equilibrium has a homoclinic orbit that lies in the boundary of its two-dimensional unstable

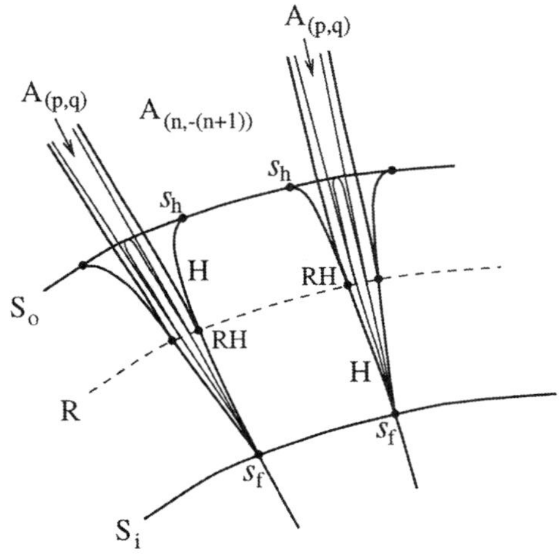


FIGURE 3. The global bifurcations bounding Arnold tongues are shown. The Arnold tongues of homotopy type $(n, -(n+1))$ extend to a segment of the curve $S_i$. Curves of homoclinic bifurcations are labelled by $H$, resonant homoclinic cycle points are denoted $RH$, points of saddle-node loop bifurcations by $s_h$, and saddle-node fan points by $s_f$. Each saddle-node fan point lies at the end of an infinite number of Arnold tongues.

manifold. Along $S_o$ between the points $s_h$, the flows have a single periodic orbit which is stable. Emanating from the points $s_h$ are curves of homoclinic bifurcations that extend into the interior of the resonance region. These homoclinic curves terminate on the saddle-node curve $S_i$ at *saddle-node fan* points $s_f$ (Baesens et al. [**4**] p. 432). Each of these complex codimension two bifurcation points $s_f$ lies at the end of homoclinic bifurcation curves of a countable number of Arnold tongues $A_{(p,q)}$. The saddle-node fans also lie at the endpoints of heteroclinic bifurcation curves that are described below. For tongues $A_{(p,q)}$ with $-q > p+1$, both homoclinic curves bounding the tongue terminate at the same saddle-node fan point. The homoclinic curves of the tongues $A_{(n,-(n+1))}$ end at adjacent saddle-node fan points. No periodic orbits of homotopy type $(p,q)$, $|pq| > 1$, are found in the interior of $S_i$. The points where the homoclinic curves cross the resonant saddle curve $R$ are resonant homoclinic bifurcation points. At each of these a bifurcation curve of saddle-nodes of periodic orbits meets the homoclinic curve and terminates. In the phase space, the periodic orbits emerging from the homoclinic orbits are unstable outside the resonant saddle curve $R$, while those emerging from the homoclinic curves inside $R$ are stable. The Arnold tongues $A_{(p,q)}$ are separated by flows with quasiperiodic minimal sets. Outside the curve $S_o$, these flows have dense trajectories. In the annulus between $S_o$ and $S_i$, the flows are Cherry flows (see Baesens et al. [**4**]) with minimal sets that are solenoids.

We next describe the additional bifurcations associated with the Arnold tongue $A_{(1,-1)}$ of homotopy type $(1,-1)$. This resonance region contains codimension one heteroclinic bifurcations and codimension two heteroclinic cycles that appear inside the region bounded by $S_i$. For parameters that lie on the $b$ axis, the systems (1) have a reflection symmetry. Figure 4 shows a collection of four phase portraits

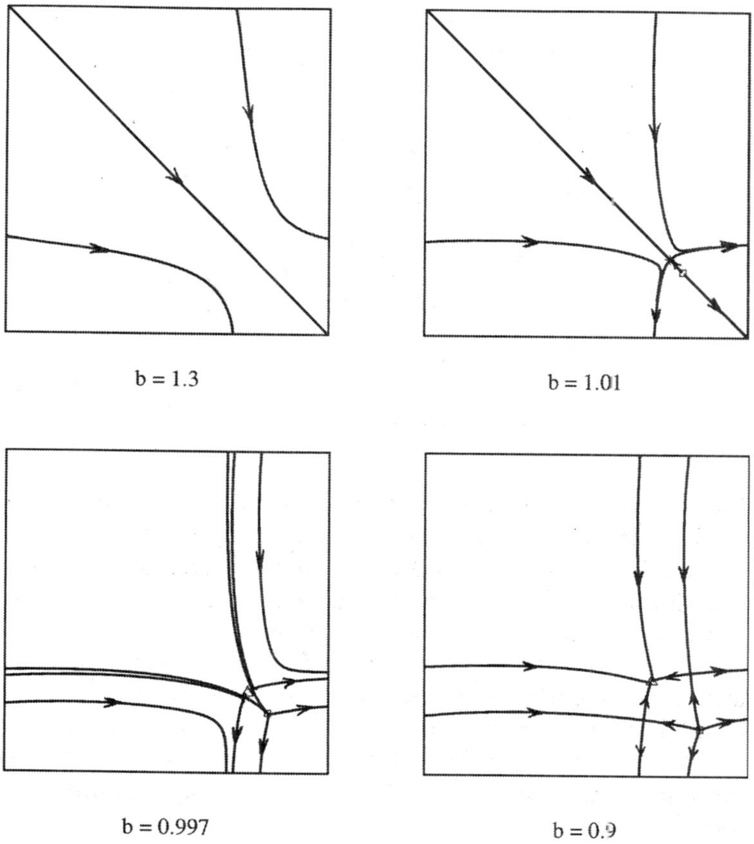


FIGURE 4. Phase portraits for symmetric vector fields with $c = 1$, $d = 0.1$ and $a = 0$ are shown for the values of $b$ displayed below the phase portraits. The parameter $b = 1.3$ lies outside the resonance region and there are two periodic orbits. The parameter $b = 1.01$ lies between $S_o$ and $S_i$. There is a source and a saddle on the invariant line and a stable periodic orbit. When $b = 0.997$, there are four equilibrium points and a stable periodic orbit. As $b$ decreases, the periodic orbit disappears when the two saddle points have a heteroclinic cycle. The value $b = 0.9$ is below this bifurcation, and all trajectories tend to equilibrium points.

for decreasing values of $b$ along this axis. Outside $S_o$, there are a pair of periodic orbits, one of which is the invariant circle $x_1 + x_2 = 2\pi$. The periodic orbits that emerge at $S_o$ lie on this invariant circle and destroy the unstable periodic orbit. The intersection of $S_i$ with the $b$ axis is a cusp at which a pair of asymmetric saddle points $p_1$ and $p_2$ emerge from the invariant line, leaving behind a sink on the axis. At a parameter value $b = b_{\text{het}}$ below the cusp point, $p_1$ and $p_2$ have a heteroclinic cycle with vertices at $p_1$ and $p_2$. The periodic orbit that exists for $b > b_{\text{het}}$ disappears at $b = b_{\text{het}}$. For $b_{\text{het}} > b > 0$, there are no periodic orbits.

The parameter regions inside the cusp of $S_i$ on the positive $b$ axis contain sequences of heteroclinic bifurcations associated with $A_{(1,-1)}$. Refer to Figure 5 for

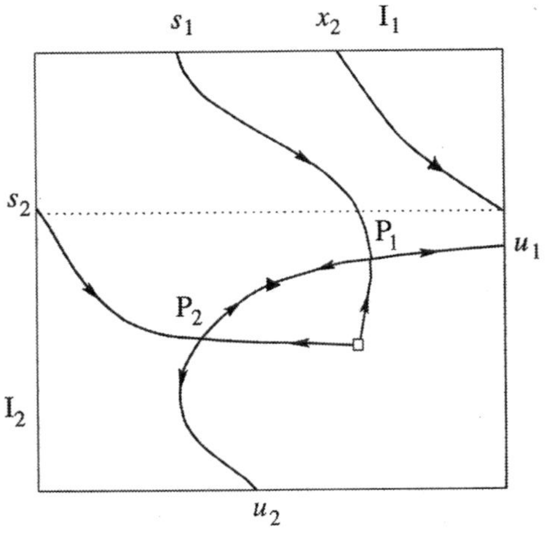


FIGURE 5. The phase portrait of the torus for parameter values inside the curve $S_i$. There are four equilibrium points, two of which are saddles labelled $p_1$ and $p_2$. The unstable manifold of $p_1$ has a branch that intersects the right-hand side of the square at $u_1$ and a branch of its stable manifold that intersects the top of the square at $s_1$. The saddle $p_2$ has a branch of its unstable manifold that intersects the bottom of the square at $u_2$ and a branch of its stable manifold that intersects the left side of the square at $s_2$. If $s_2$ lies above $u_1$, we continue the stable manifold of $p_2$ until it intersects the top of the square at $x_2$. The segment $(s_1, x_2)$ returns to the top of square.

a picture of the phase space for parameters in this region. We use the unit square as a fundamental domain for $T^2$. There are four equilibrium points, a sink, a source and two saddles. The two saddle points are labeled $p_1$ and $p_2$. Each saddle has one stable separatrix that goes to the sink without intersecting the boundary of the unit square and one unstable separatrix that comes from the source without intersecting the boundary of the unit square. The other separatrices of the saddles $p_1$ and $p_2$ do intersect the boundary of the unit square in points $s_1, s_2, u_1, u_2$. Points to the left of $s_1$ and above $s_2$ tend to the sink and those below $u_1$ and to the right of $u_2$ come from the source. We need to follow points to the right of $s_1$ and below $s_2$ to determine the remainder of the phase portrait of these fields.

Let $I_1 = (s_1, 1)$ be the interval on the top of the unit square to the right of $s_1$, and let $I_2 = (0, s_2)$ be the interval on the left edge of the square below $s_2$. We examine the return map of $I_1$ to the top of the unit square. Points that intersect the left side of the square above $s_2$ tend to the sink and do not return. Thus, if $u_1 > s_2$, there is no recurrence for the map and all points have $\alpha$- and $\omega$-limit sets that are equilibria. When $u_1 < s_2$ and $s_1 < u_2$, there is an interval $(s_1, x_2) = J \subset I_1$ that does return to the top of the square with $x_2$ denoting the first point on the backwards trajectory of $s_2$ to intersect the top of the square. Let $\theta$ be the first return map to the top of the square. The map $\theta$ is monotone increasing and has a continuous extension to the closed interval $[s_1, x_2]$. The derivative of $\theta$ appears to be smaller than one in our numerical studies, so we assume that $\theta' < 1$ in the following discussion. There are three cases to consider:

1. $\theta(J) \subset J$,
2. $s_1 \in \theta(J)$,
3. $\theta(J) \subset (0, s_1)$.

In the first of these cases, there is a fixed point that is the intersection of $J$ with a periodic orbit of homotopy type $(1, -1)$. In the third case, all of the points of $J$ tend to the sink after a single return to the top of the square. The second case is more complicated than the first and third. All points of $J$ either tend to the sink or lie in the stable manifold of $p_1$, but they may make repeated returns to $J$. If $\theta^n(x_2) = s_1$, there is a heteroclinic orbit that intersects the top of the square in $n+1$ points. As we proceed along the interior of the inner saddle-node curve $S_i$ towards the cusp with negative values of $a$ increasing towards 0, we move from the third case described above to the second and then to the first. We encounter a sequence of heteroclinic bifurcations which accumulate at parameters for which $\theta(s_1) = s_1$. These parameter values are ones for which the point $p_1$ has a homoclinic orbit, and they bound the Arnold tongue $A_{(1,-1)}$. All of these curves of heteroclinic and homoclinic bifurcations meet at the codimension two point $(a, b) = (0, b_{\text{het}})$ of the heteroclinic cycle. The bifurcation curves to the right of the $b$ axis are reflections of the ones we have described to the left of the $b$ axis.

We want to connect our analysis of the global bifurcations inside the saddle-node curve $S_i$ with that for the annulus bounded by the two saddle-node curves $S_i$ and $S_o$. The end points of the heteroclinic bifurcations on $S_i$ are the codimension two saddle-node fan points described above. At these parameters, the branch of the stable manifold of $p_1$ passing through $s_1$ coincides with the unstable manifold of the saddle-node point $p_2$. We can extend these connections to closed curves by adjoining either branch of the unstable manifold of $p_1$. The two closed curves formed by adjoining one of the branches of the unstable manifold of $p_1$ have homotopy types of the form $(n-1, -n)$ and $(n, -(n+1))$. This saddle-node fan point is the end point of homoclinic bifurcation curves bounding $A_{(p,q)}$ for

$$\frac{p}{q} \in \left( -1 + \frac{1}{n+1}, -1 + \frac{1}{n} \right).$$

Thus the codimension two points on $S_i$ are the common endpoints of heteroclinic bifurcation curves inside $S_i$ and a countable set of homoclinic bifurcation curves lying in the annulus between $S_i$ and $S_o$. This completes our description of the bifurcation diagram inside $S_i$ near the cusp along the positive $b$ axis in the $(a, b)$ plane. Figure 6 depicts the bifurcation diagram of this region of the $(a, b)$ plane.

Our description of the bifurcations of two-parameter families (1) with varying $a$ and $b$ and $|c/d|$ large contains only generic codimension one and two bifurcations. We conjecture that this description is not only consistent but also correct. In particular, we believe that it includes all of the bifurcations that occur in these families. The unfoldings of each of the codimension two bifurcations have been described previously in the literature. The only novelty is the appearance of the countable sequences of heteroclinic bifurcations that meet at the heteroclinic cycles. It is evident that there are significant changes in the global bifurcation diagrams as $|c/d|$ decreases, but we have not tried to classify these or to determine which codimension three global bifurcations are present within the family (1).

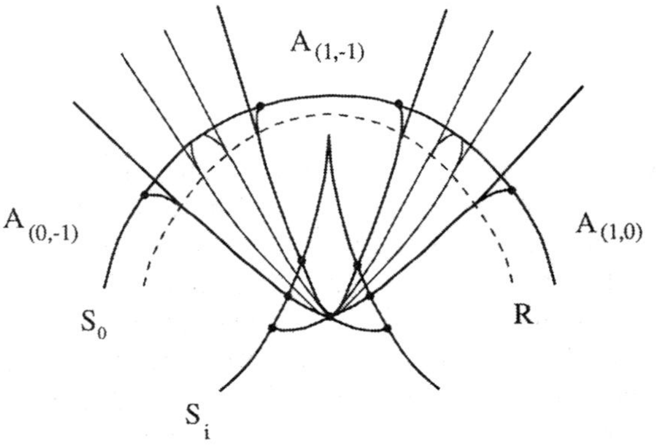


FIGURE 6. An overview of global bifurcations in the vicinity of the positive $b$ axis. The three Arnold tongues with homotopy types $(0,-1)$, $(1,-1)$ and $(1,0)$ are large, with the tongue $A_{(1,-1)}$ extending into the interior of $S_i$. The codimension two bifurcation points along $S_o$ are transversal intersections of saddle-nodes of equilibriums and limit cycles and saddle-node loop points. The codimension two bifurcation points on the curve of resonant saddles denoted $R$ are resonant homoclinic bifurcations. The bifurcations on $S_i$ are saddle-node fan points at which infinite collections of Arnold tongues terminate. Inside $S_i$ are curves of heteroclinic bifurcations that all meet at the heteroclinic cycle point on the $b$ axis.

## References


[1] V. I. Arnold, *Small divisors. I. On mappings of the circle onto itself*, Izv. Akad. Nauk SSSR, Ser. Mat. **25** (1965), 21–86; English transl., Amer. Math. Soc. Transl. Ser. 2, vol. 46, Amer. Math. Soc., Providence, RI, 1995, pp. 213–284.

[2] Y. Braiman, T. A. B. Kennedy, K. Wiesenfeld, and A. I. Khibnik, *Entrainment of solid-state laser arrays*, Phys. Rev. A **52** (1995), 1500–1506.

[3] C. Baesens, J. Guckenheimer, S. Kim, and R. MacKay, *Simple Resonance Regions of Torus Diffeomorphisms, Patterns and Dynamics of Reactive Media* (Aris, Aronson and Swinney eds.), IMA Volumes in Mathematics and its Applications, vol. 37, Springer-Verlag, New York, 1991, pp. 1–9.

[4] ———, *Three coupled oscillators: mode-locking, global bifurcations and toroidal chaos*, Phys. D **49** (1991), 387–485.

[5] J. Guckenheimer and P. J. Holmes, *Nonlinear oscillations, dynamical systems, and bifurcations of vector fields*, Springer-Verlag, Berlin, New York, 1983.

[6] M. R. Herman, *Sur la conjugaison differentiable des difféomorphismes du cercle à des rotations*, Publ. Math IHES **49** (1979), 5–234.

[7] A. I. Khibnik, Y. Braiman, T. A. B. Kennedy, and K. Wiesenfeld, *Phase analysis of two lasers with injected field*, Phys. D. (to appear)

[8] J. C. Yoccoz, *Conjugaison différentiable des difféomorphismes du cercle dont le nombre de rotation vérifie une condition diophantienne*, Ann. Sci. École Norm. Sup. (4) **17** (1984), 333–359.


DEPARTMENT OF MATHEMATICS, CORNELL UNIVERSITY, ITHACA, NEW YORK, 14853

CORNELL THEORY CENTER AND THE INSTITUTE OF MATHEMATICAL PROBLEMS IN BIOLOGY, PUSHCHINO, RUSSIA

Translated by THE AUTHORS

Amer. Math. Soc. Transl.
(2) Vol. **200**, 2000

# Isoenergetical Structure of Integrable Hamiltonian Systems in an Extended Neighborhood of a Simple Singular Point: Three Degrees of Freedom

L. M. Lerman

*To the memory of Professor E. A. Leontovich-Andronova*

ABSTRACT. An integrable Hamiltonian vector field with three degrees of freedom (3D IHVF) in a saturated neighborhood (with respect to the related Poisson action) of its simple singular point is studied. Points of the following type are considered: elliptic-hyperbolic (2 pairs of pure imaginary eigenvalues and 1 pair of nonzero real ones), csf-points (1 pair of pure imaginary eigenvalues and the quadruple of complex ones), cs-points (1 pair of pure imaginary eigenvalues and 2 pairs of real ones), and sfs-points (1 pair of the reals and the quadruple of complex ones). Corresponding invariants of isoenergetic equivalence are found and examples where families of nonorientable saddle invariant two-dimensional tori exist are constructed.

## 1. Introduction

The study of the structure of integrable Hamiltonian vector fields (IVFs) now constitutes a rather developed chapter of the theory of dynamical systems and symplectic geometry (see, for instance, [**1, 2, 3, 4, 5**] and references therein). But these works mainly deal with 2D IVFs, or with selected multidimensional problems. As far as general multidimensional IVFs are concerned, especially from the viewpoint of their isoenergetical structure, the results obtained are more modest. But this is the main point of interest when one considers some IVF that arises in applications.

The goal of this paper is to present possible types of isoenergetic orbit behavior for a Hamiltonian integrable vector field (IVF, below) with three degrees of freedom in an extended neighborhood of its simple singular point. To be precise, let us recall several notions and definitions.

Let $(M, \Omega)$ be a $C^\infty$-smooth $2n$-dimensional symplectic manifold with symplectic 2-form $\Omega$ [**6**], and $H$ a smooth ($C^\infty$) Hamiltonian with the corresponding

---

2000 *Mathematics Subject Classification.* Primary 37J20; Secondary 37Jxx.

The author acknowledges the support of the Russian Foundation for Basic Research under the grant 98-01-00795, INCAS under the grant 98-1-03, and the program "Universities of Russia" (project No. 1905). The paper was completed during the author's stay at the Free University of Berlin. The author is grateful to Professor B. Fiedler for the invitation and warm hospitality.

vector field $X_H$. This vector field is said to be (Liouville) integrable if there exist $n-1$ additional smooth integrals $F_1, \dots, F_{n-1}$ such that the set $\{H, F_1, \dots, F_{n-1}\}$ is involutive, and the differentials $dH, dF_1, \dots, dF_{n-1}$ are independent in an open dense subset of $M$. From now on we suppose that the Hamiltonian vector fields $X_H, X_{F_1}, \dots, X_{F_{n-1}}$ are complete, that is, generate actions of the group $\mathbb{R}$ (each orbit is extendable to $\mathbb{R}$). Below we denote by $(X_H, F)$, $F = (F_1, \dots, F_{n-1})$, an IVF with the Hamiltonian $H$ and additional integrals $F = (F_1, \dots, F_{n-1})$. The IVF generates a smooth Poisson action $\Phi$ of the group $\mathbb{R}^n$ on $M$ [**6**]. The orbits of this action are immersions of $\mathbb{R}^p \times T^q$, $p + q \leqslant n$, where $T^q$ is the standard $q$-dimensional torus $\mathbb{R}^q/\mathbb{Z}^q$. The orbits with $p+q=n$ are called *nonsingular*, those with $p+q<n$ are *singular*. The Liouville–Arnold theorem states that a compact connected nondegenerate joint level of all integrals is an $n$-dimensional torus, but if this set contains singular orbits (*singular leaf*) its structure can be topologically rather complicated. We suppose that $\Phi$ is a Lagrange stable action, that is, the closure of any of its orbits is a compact set. Singular points are zero-dimensional orbits. To have a singular point $p$ of $\Phi$, it is sufficient for $p$ to be an isolated singular point of $X_H$ or any $X_{F_i}$.

DEFINITION 1 ([**12**]). A singular point $p$ of the generated action $\Phi$ is called *simple* if the commutative algebra of quadratic functions on $T_pM$ with respect to the Poisson bracket generated by the quadratic parts of the functions $H, F_1, \dots, F_{n-1}$ at $p$ is Cartanian.

REMARK 1. Another more "practical" definition is the following. Let $A_1, \dots, A_n$ be the linear symplectic operators given by linearizing the vector fields $X_H, X_{F_1}, \dots, X_{F_{n-1}}$ at $p$. These operators commute and the algebra is the linear span of $A_1, \dots, A_n$ with complex coefficients and the commutator of two matrices as the operation. This algebra is required to contain an operator with simple eigenvalues.

When one deals with an IVF, the first task is to study the related Liouville foliation generated by $\Phi$ [**7**], since any of its orbits is an invariant set of all the fields $X_H, X_{F_1}, \dots, X_{F_{n-1}}$. This means that one needs to define when two IVFs will be considered equivalent. The simplest natural equivalence relation for two IVFs is topological equivalence [**7**]. To formulate this notion, let us call a set $S \subset M$ *saturated* with respect to the action if it consists of complete orbits. Let $(X_H, F)$ and $(X_K, G)$ be two IVFs and $\Phi$, $\Psi$ the corresponding actions.

DEFINITION 2. The IVFs $(X_H, F)$ and $(X_K, G)$ are called *topologically equivalent* in their saturated subsets $S_1$ and $S_2$ if there is a homeomorphism $h\colon S_1 \to S_2$ such that $h$ sends orbits of the action $\Phi$ onto orbits of $\Psi$.

If, instead of $h$ being a homeomorphism, it is a diffeomorphism, then one speaks of differential equivalence, etc.

Studying IHVs has two aspects. One aspect is the study of IVF for the purposes of symplectic geometry and topology, for instance, in order to understand the structure of symplectic manifolds, the geometry of symplectomorphisms, etc., [**26**]. Another aspect is directed to studying IVFs as the first (or even the zeroth) step towards understanding the global structure of (nonintegrable) Hamiltonian systems. From this latter viewpoint, topological equivalence is not sufficient, since one needs to know what is the behavior of the flow generated by $X_H$, and, because the levels $M_h = \{H = h\}$ are invariant sets, one needs to follow how the dynamics changes with respect to $h$, the main parameter of the system. Thus, another equivalence relation is more appropriate.

DEFINITION 3 ([**8**]). Two IVFs $(X_H, F)$ and $(X_K, G)$ are called *isoenergetically equivalent* if there is a topological equivalence $g\colon M \to M$ respecting the fibration by $\{H = h\}$, that is, $g(M_h) \subset M_k$, and vice versa, with $M_h = \{H = h\}$ and $M_k = \{K = k\}$.

REMARK 2. In principle, we are interested in the behavior of trajectories of the vector field $X_H$, so that studying the orbit foliation generated by $\Phi$ is an auxiliary (though important) problem. This implies that one may change the additional integrals $F_1, F_2, \dots, F_{n-1}$ if this does not change the orbit foliation. In particular, this foliation does not change if the condition of density of the Liouville tori with incommensurate frequency vectors is fulfilled (see, for instance, [**26**]). When investigating the structure of $X_H$ in an extended neighborhood of a simple singular point, this condition always holds, and so, one may change integrals if necessary. It follows from this remark that we may suppose, without loss of generality, that restrictions of $H$ and other integrals to some invariant symplectic disks in a neighborhood of a simple singular point $p$ are also simple. We use this fact further without special mention.

In the case of two-degrees-of-freedom IVFs, the isoenergetical classification and its invariants for an IVF with a Bottian $H$ restricted to a nondegenerate ($dH \neq 0$) compact level of $H$ were given in [**10**] and [**11**]. When investigating an IVF in the large, one needs, in addition, to weaken the restriction on $H$ to be Bottian to examine the isoenergetical structure of the IVF in an extended neighborhood of singular points (recall that an extended neighborhood of a singular point is an open set which is saturated with respect to the action and to which all orbits containing $p$ in their closure belong. This was done in [**12, 13, 14**] (in more detail this approach and results are presented in [**5**]) for the case in which one singular point belongs to a singular leaf, and then was extended to the case of several such points in [**18, 15, 16**].

An important next step was carried out in [**19**], where it was shown that for an extended neighborhood of a simple singular leaf of $n$-degrees-of-freedom IVF there exists a covering of this neighborhood such that the Liouville foliation in the covering space becomes topologically equivalent to the direct product of extended neighborhoods of simple singular points of a 1D IVF and/or 2D IVF (for saddle-foci). This result, however, does not provide conditions of isoenergetical equivalence. It is worth emphasizing that the results of [**19**] are not sufficient to comprehend the structure of an IVF in the large, since only simple singular orbits have been studied. In order to understand the structure in the large, one needs, in addition, to investigate the bifurcations of IVFs, especially if we are concerned with the multidimensional case $n \geqslant 3$. To explain how these bifurcation problems naturally arise, let us recall some results from [**7, 9**].

Let $x$ be a point of a singular orbit $O_x$, and $k \in \mathbb{N}$ be the number of independent differentials among $dH, dF_1, \dots, dF_{n-1}$ at the point $x$ ($k$ is equal to the orbit dimension). The reduction procedure [**20**], being applied locally near $x$, generically yields (see [**7, 9**]) a $k$-parametric smooth family of reduced IVF with $n-k$ degrees of freedom in a neighborhood of the singular point that corresponds to $O_x$. For the case in which this point is simple, the singular point can be included in a $k$-parameter family of such singular points, so that every IVF will have a simple singular point of the same type. For instance, if $n = 2$ and $k = 1$, we obtain two types of reduced 1D Hamiltonians with a singular point of the center or saddle type. But if the reduced system has a nonsimple singular point, bifurcations of the point type along the family of singular points are possible. For $n = 2$ it is (generically) a bifurcation through a parabolic point [**7**]. For $n = 3$, the situation becomes richer. Near a one-dimensional orbit, we come to a one-parameter family of 2D IVFs, and if the reference singular point is not simple, one can generically obtain the bifurcations: "elliptic point–saddle-focus", "elliptic point–saddle-center", "saddle-center–saddle", and "saddle–saddle-focus". The first three bifurcations occur in nonintegrable Hamiltonian systems near a singular point at resonance as well. Moreover, the related normal forms are integrable, so these bifurcations of the normal form are the same. The bifurcations of reduced phase portraits can be found, for instance, in [**22**]. As will be shown below, the latter bifurcation "saddle–saddle-focus" is locally a bifurcation of the Liouville foliation, but not a bifurcation of the related Hamiltonian vector field. It becomes a bifurcation of the Hamiltonian system if one considers an extended neighborhood of the point (see Section 2).

In a 3D IVF, besides one-dimensional singular orbits, two-dimensional ones may exist. Near a point on a two-dimensional orbit, two-dimensional orbits generically form a two-parameter family, and reduction provides a two-parameter family of 1D Hamiltonians that unfolds the Hamiltonian that has a singular point. Thus, generically, in such a family of functions codimension 2 singular points can exist. These singularities can be of two types, depending on the sign of some coefficient $\alpha$ in the normal form of the third order. The related phase portraits are borrowed from [**23**] and plotted in Figures 1, 2.

The paper is organized as follows. In Section 2 we consider an example of the reconstruction of a Liouville foliation as we pass through a nonsimple periodic orbit in a 3D IVF. In Sections 3, 4, 5, 6 the main points determining the isoenergetic structure, and related classification theorems are presented. As usual, the most complex cases are extended neighborhoods of pure hyperbolic points, sfs-points, and s-points. In Section 7 we only describe the principal points which lead to this complexity and present an example of an IVF with sfs-point and nonorientable saddle 2-tori. Because of the restriction of space in this volume, all proofs are omitted. We hope to publish them elsewhere, together with a detailed study of saddle cases.

## 2. Foliation reconstructions when passing through double real resonance in a one-parameter integrable unfolding of an IVF: an example

Let a 3D IVF $(X_H, F_1, F_2)$ be given. We assume that $X_H$ has a one-parameter family of periodic trajectories $\gamma_c$, which are one-dimensional orbits of the related Poisson action. Next, we suppose all orbits of a vector field $X_{F_2}$ to be periodic, so that reduction with respect to the integral $F_2$ leads to a one-parameter family of reduced 2D Hamiltonians $h_\varepsilon$ with additional integrals $f_\varepsilon$. Since the curves $\gamma_c$ are trajectories of $X_{F_2}$ as well, it follows that after reduction they become singular points $p_\varepsilon$.

ASSUMPTION 1. The singular point $p_0$ is nonsimple for the reduced Poisson action, the points $p_\varepsilon$, $\varepsilon < 0$, are focus-foci, the singular points $p_\varepsilon$, $\varepsilon > 0$, are saddle-saddles (see [**5**] for related definitions).

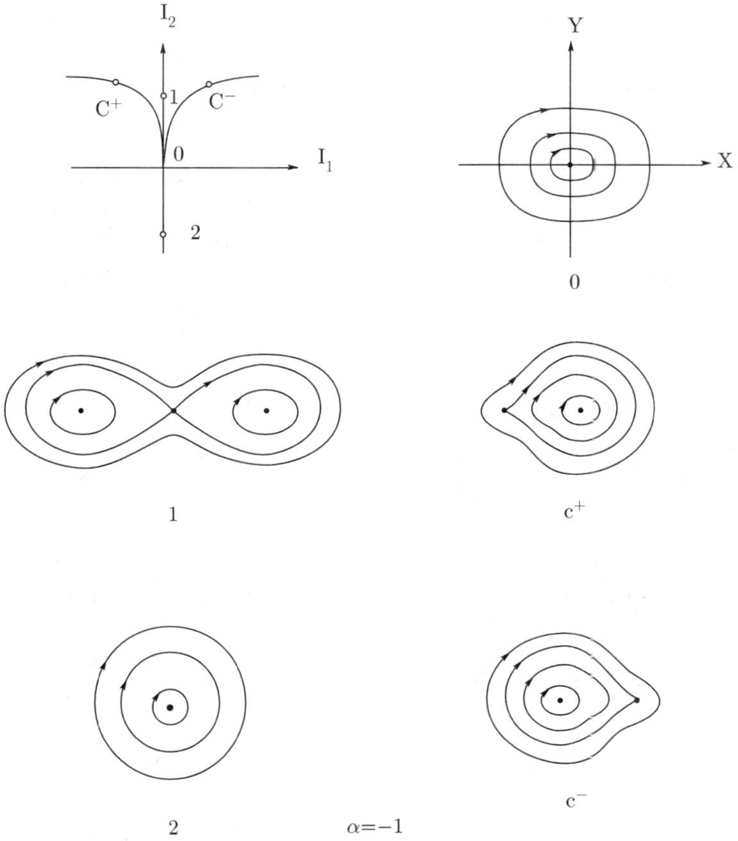


FIGURE 1

It follows, in particular, that one can assume $p_\varepsilon$ fixed, $p_\varepsilon = p$.

ASSUMPTION 2. For any $\varepsilon$ a singular leaf of $p$ is compact and contains only one singular point $p$ and does not contain closed one-dimensional orbits.

ASSUMPTION 3. There is a symplectic frame near $p$ such that $h_\varepsilon$, $f_\varepsilon$ have local representations

$$h_\varepsilon = x_1y_1 + x_2y_2 + x_1y_2 + \varepsilon x_2y_1 + Ax_2^2y_1^2, \qquad f_\varepsilon = x_1y_1 + x_2y_2.$$

It is readily verified that a singular point $O$ at the origin changes its type as is required in Assumption 1.

REMARK 3. The choice of the form of $h_\varepsilon$ is not accidental. It is simply a piece of the fourth-order normal form of any 2D Hamiltonian that undergoes a double real resonance $(\lambda, \lambda, -\lambda, -\lambda)$ with two-dimensional Jordan blocks for any $\lambda$ at the linear part of the vector field at $\varepsilon = 0$.

First let us consider the local structure of the corresponding vector field and the Poisson action. Since $O$ is of the saddle type for all $\varepsilon$, there are two smooth local mutually transverse Lagrangian disks, $W^s$, $W^u$, through $O$, its stable ($y_1 = y_2 = 0$) and unstable ($x_1 = x_2 = 0$) manifolds. These manifolds belong to the singular leaf determined by the system $h_\varepsilon = f_\varepsilon = 0$. Its solutions are not exhausted by

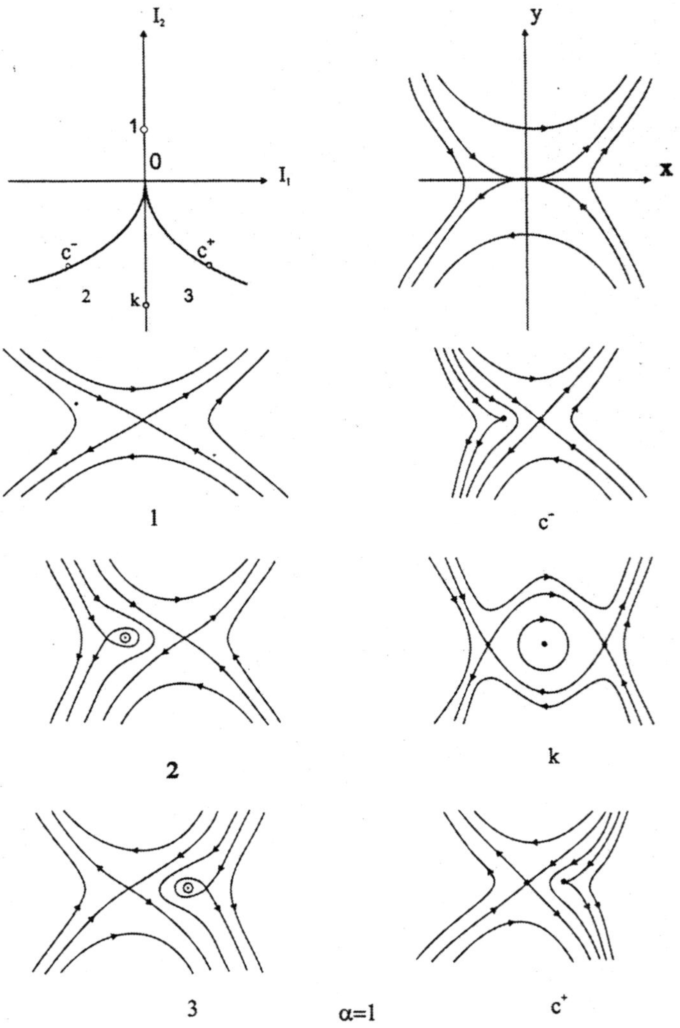


FIGURE 2

points of the stable and unstable disks, other solutions lie on the graph given by the functions $y_1 = x_1^2/Ax_2^3$, $y_2 = -x_1^3/Ax_2^4$. For the Poisson action, each disk, for instance $W^s$, consists of 5 orbits: one zero-dimensional (point $O$), two one-dimensional ($y_1 = y_2 = x_1 = 0$, $x_2 > 0$ or $x_2 < 0$), and two two-dimensional (semi-disks $y_1 = y_2 = 0$, $x_1 > 0$ or $x_1 < 0$). Let us extend each one-dimensional orbit with respect to the action. Since the limit points of a one-dimensional orbit coincide with $p$ (due to Assumption 2), these orbits have to merge upon continuing with the two local one-dimensional orbits lying on $W^u$ ($x_1 = x_2 = y_2 = 0$, $y_1 > 0$ or $y_1 < 0$). Thus, two loops $\Gamma_1$, $\Gamma_2$ are formed. The unstable manifold $W^u$ being continued near $\Gamma_i$ comes back to the stable manifold $W^s$ but does not merge with it, only touches it along a one-dimensional orbit. In order to describe the foliation structure near loops and its reconstructions when the parameters $\varepsilon$, $h$ change, we take two cross-sections to the orbits, one of them given by the equation $x_2 = d$,

$d > 0$, and the other by $x_2 = -d$. Every section is foliated by levels $h_\varepsilon = h$ into two-dimensional curvilinear rectangles given by the inequalities $|x| \leqslant d/2$, $|y_1| \leqslant \delta$. Each rectangle is the graph of two functions $h_\varepsilon = h$, $x_2 = d$, or $x_2 = -d$. We describe the foliation on the first section and explain what the changes on the other section are: the Liouville foliation is transverse to the section, so it generates a one-dimensional foliation on each rectangle, this foliation is into level lines of the restriction of the function $f_\varepsilon$ to the rectangle. The function thus obtained will be denoted by $k(x, y)$ (we omit subindices).

First, we describe reconstructions on the disk $h = 0$ when $\varepsilon$ changes. Then one obtains $k = y(x^2 - \varepsilon d^2 - Ad^3 y)$. An easy but tedious calculation of critical points and their types, as well as values of $k$ at these points, give the corresponding foliation plotted in Figure 3. It is evident that singular points of the foliation, besides those on the axis $y = 0$ (the trace of $W^s$), correspond to periodic orbits. Singular points with $y = 0$ correspond to homoclinic orbits to $p$. For $\varepsilon = 0$ the critical point $x = y = 0$ is a degenerate saddle (its index is equal to 2), therefore it should bifurcate under varying parameters. This can be seen from Figure 3 (along the line $h = 0$). Thus we observe that the passage from saddle to saddle-focus is accompanied by the connection formed by two heteroclinic orbits collapsing into an elliptic orbit. The same foliation is generated on the section $x_2 = -d$, moreover, one foliation transforms to the other one under the map $(x, y) \to -(x, y)$.

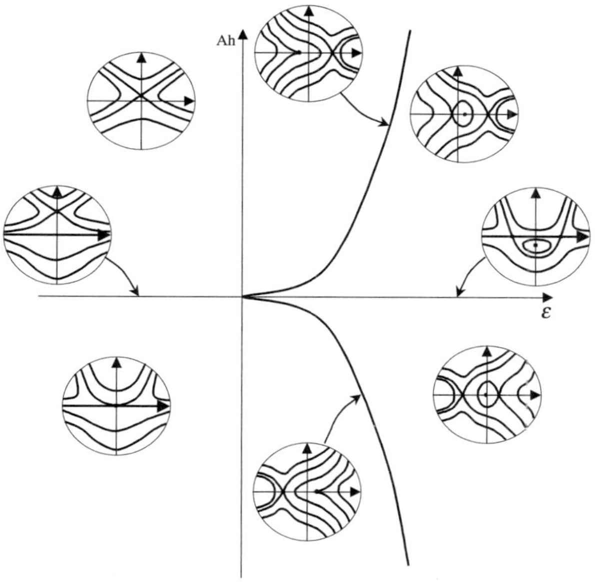


FIGURE 3

Now let us trace back the reconstructions in the two-parameter unfolding. We shall place ourselves on the level $h =$ const and vary $\varepsilon$. The restriction of the function $f_\varepsilon$ to this level takes the form

$$k = \frac{hd + y(x^2 - \varepsilon d^2 - Ad^3 y)}{x + d}.$$

Critical points of this function are determined by the equations

$$3x^4 + 4dx^3 - 2\varepsilon d^2x^2 - 4\varepsilon d^3x - \varepsilon^2 d^4 = 4Ad^4h, \qquad y = \frac{x^2 - \varepsilon d^2}{2Ad^3}.$$

The first equation with respect to the new variable $\xi = x/d$ can be rewritten as

$$3\xi^4 + 4\xi^3 - 4Ah = 2\varepsilon(\xi^2 + 2\xi + \varepsilon/2).$$

Its roots close to zero can be easily determined, the curve corresponding to multiple roots on the plane $(\varepsilon, h)$ is given by the relation $h = (\pm\sqrt{3}\varepsilon^{3/2} - 2\varepsilon^2)/6A$. This curve (except for its cusp) corresponds to the emergence of a cusp (parabolic) critical point. The related periodic orbit is of parabolic type (see [**7**]). It is clear from this that for $\varepsilon < 0$ there is only one critical point, the saddle, and for $\varepsilon > 0$ a bifurcation takes place. The related foliations are depicted in Figure 3.

## 3. 3D IVFs, bifurcation diagrams

Now let us turn to the study of the behavior of 3D IVF in an extended neighborhood of a simple singular point $p$. Without loss of generality we assume that $H(p) = F_1(p) = F_2(p) = 0$. Recall that, due to the result of Vey–Eliasson [**24, 25**], there is a symplectic frame near $p$ such that the Hamiltonian $H$ and additional integrals $F_1$, $F_2$ written in the form

(1) $$H = h(\xi_1, \xi_2, \xi_3), \quad F_1 = f_1(\xi_1, \xi_2, \xi_3), \quad F_1 = f_1(\xi_1, \xi_2, \xi_3),$$

where, depending on the type of the point, quadratic homogeneous polynomials $\xi_1$, $\xi_2$, $\xi_3$ have the following form:

1. $\xi_i = (x_i^2 + y_i^2)/2$, $i = 1, 2, 3$ (elliptic point, or briefly, e-point);
2. $\xi_i = (x_i^2 + y_i^2)/2$, $i = 1, 2$, $\xi_3 = x_3y_3$ (elliptic-hyperbolic point, eh-point);
3. $\xi_1 = (x_1^2 + y_1^2)/2$, $\xi_2 = x_2y_2 + x_3y_3$, $\xi_3 = x_2y_3 - x_3y_2$ (center-saddle-focus, csf-point);
4. $\xi_1 = (x_1^2 + y_1^2)/2$, $\xi_j = x_jy_j$, $j = 1, 2$ (center-saddle, cs-point);
5. $\xi_1 = x_1y_1 + x_2y_2$, $\xi_2 = x_1y_2 - x_2y_1$, $\xi_3 = x_3y_3$ (saddle-focus-saddle, sfs-point);
6. $\xi_i = x_iy_i$, $i = 1, 2, 3$ (saddle, s-point).

Thus, a 3D IVF can have 6 types of simple singular points. In these coordinates the condition of simplicity merely means that the determinant composed of the coefficients in the linear parts with respect to $\xi_1$, $\xi_2$, $\xi_3$ of the functions $h$, $f_1$, $f_2$ does not vanish. As a corollary of the normal form, the next statement is valid [**9**].

THEOREM 1. *There is a neighborhood $U$ of a simple singular point in $M$ and a neighborhood $V$ of the origin in $T_pM$ such that the Liouville foliations generated by $H$, $F_1$, $F_2$ in $U$ and $\xi_1$, $\xi_2$, $\xi_3$ in $V$ are smoothly equivalent.*

Thus, only the global behavior in some saturated set is a nontrivial problem for IVFs.

Now let us pass to the study of the isoenergetical structure of the IVF in extended neighborhoods of simple singular points. Recall that the case of an elliptic point was studied (for any $n$) in [**21**]. Applying this result to our case $n = 3$, we can state

THEOREM 2. *A complete isoenergetical invariant of an IVF in a sufficiently small saturated neighborhood of an elliptic singular point is the absolute value of the signature of the quadratic form given by the Hamiltonian at the singular point.*

Therefore, we must only consider the remaining 5 types. To simplify our considerations, we shall assume that

(1) the connected component of the set $H = F_1 = F_2 = 0$ (singular leaf of the Liouville foliation), except for $p$, does not contain other singular points;
(2) the boundary of any orbit (closure of the orbit minus the orbit itself), being a saturated set, consists of orbits of lesser dimension than the dimension of the orbit itself.

Further on we always assume that for $X_H$ the coefficients at $\xi_1$, $\xi_2$, $\xi_3$ in the linear part of $h$ do not vanish.

Below we use the term *orbit* when we deal with action orbits, and the term *trajectory* when we deals with orbits of the specific vector fields (mostly, $X_H$).

To construct extended neighborhoods and study orbit behavior, we need to describe bifurcation diagrams for each type of simple singular points. By a bifurcation diagram $B$ we mean the image with respect to the momentum mapping $\mu$ of the set of singular orbits [**28**]. Since an extended neighborhood is saturated and all action orbits in it pass through the local neighborhood $U$, where the Vey–Eliasson theorem holds, the bifurcation diagram can be constructed by means of the local Hamiltonian and additional integrals can be written in normal form (1). Due to simplicity of singular points under consideration, all bifurcation diagrams are diffeomorphic to those constructed via the quadratic functions $\xi_1$, $\xi_2$, $\xi_3$ for each case. In this section by the modified momentum mapping we mean the mapping $\kappa\colon (x, y) \to (\xi_1, \xi_2, \xi_3)$ with the related functions $\xi_i$ of the preceding section. It is evident that the genuine momentum mapping $\mu$ is the composition of $\kappa$ and $\nu\colon (\xi_1, \xi_2, \xi_3) \to (H, F_1, F_2)$, which gives a diffeomorphic change of the bifurcation diagram only.

**1. eh-point.** Let us denote

$$H = h(\xi_1, \xi_2, \xi_3) = \omega_1\xi_1 + \omega_2\xi_2 + \lambda\xi_3 + \cdots .$$

Since $\xi_1 \geqslant 0$, $\xi_2 \geqslant 0$, the modified momentum mapping $\kappa$ takes values in the region $\xi_1 \geqslant 0$, $\xi_2 \geqslant 0$, $|\xi_3| \leqslant d$. In $U$ there are three four-dimensional invariant symplectic disks filled with two-dimensional orbits:

$$\Sigma_4^1 : x_1 = y_1 = 0, \qquad \Sigma_4^2 : x_2 = y_2 = 0, \qquad \Sigma_4^3 : x_3 = y_3 = 0,$$

and three two-dimensional symplectic disks filled with one-dimensional orbits:

$$\Sigma_2^3 : x_1 = y_1 = x_2 = y_2 = 0, \qquad \Sigma_2^2 : x_1 = y_1 = x_3 = y_3 = 0,$$
$$\Sigma_2^1 : x_2 = y_2 = x_3 = y_3 = 0.$$

Any two-dimensional disk is the transverse intersection of two four-dimensional ones. For $\sigma$ we choose the region $\xi_1 + \xi_2 \leqslant \rho$, $|\xi_3| \leqslant d$. What is important for the following is the position of the disk $h = 0$ in $\sigma$. Depending on whether $\omega_1\omega_2 > 0$ or $\omega_1\omega_2 < 0$, this disk belongs to the unique quadrant $\xi_1 \geqslant 0$, $\xi_2 \geqslant 0$, $\xi_3 \geqslant 0$ ($\xi_1 \geqslant 0$, $\xi_2 \geqslant 0$, $\xi_3 \leqslant 0$), or it transversely intersects the disk $\xi_3 = 0$. The faces $\xi_1 = 0$, $\xi_2 = 0$ and triangle $\xi_3 = 0$ in $\sigma$ are images of four-dimensional symplectic disks, and the segments $\xi_1 = \xi_2 = 0$, $\xi_1 = \xi_3 = 0$, $\xi_2 = \xi_3 = 0$ are the images under $\kappa$ of two-dimensional symplectic disks.

It is evident that the cases $\omega_1\omega_2 > 0$ and $\omega_1\omega_2 < 0$ lead to isoenergetically different systems since there are no two-dimensional orbits lying in the level $H = 0$ in the first case and there is a one-parameter family of such orbits in the second case (in $\sigma$ they correspond to the intersection curve of $H = 0$ and $\xi_3 = 0$). The relevant bifurcation diagrams and the image of an extended neighborhood $V$ are depicted in Figure 4.

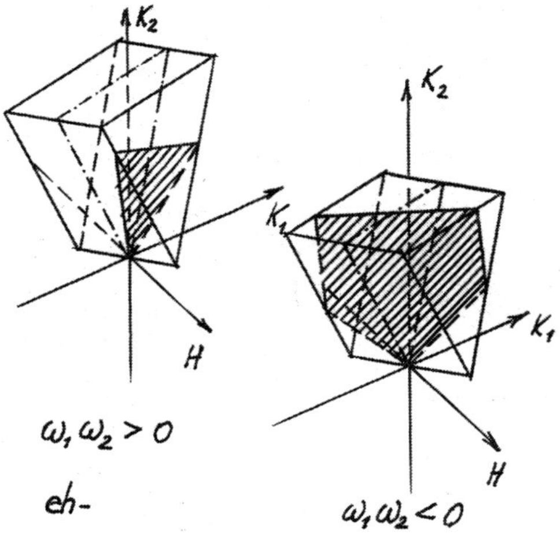


FIGURE 4

**2. csf-point.** Here

$$H = h(\xi_1, \xi_2, \xi_3) = \omega\xi_1 + \alpha\xi_2 + \beta\xi_3 + \dots, \qquad \omega > 0,\ \alpha > 0.$$

In this case only one four-dimensional symplectic invariant disk $\Sigma_4$: $x_1 = y_1 = 0$ exists, and one two-dimensional disk $\Sigma_2$: $x_2 = y_2 = x_3 = y_3 = 0$. Since the surfaces $h = 0$ and $\xi_1 = 0$ are transverse, for $\sigma$ we can take the region given by relations $|h| \leqslant \varepsilon$, $0 \leqslant \xi_1 \leqslant d$, $\xi_2^2 + \xi_3^2 \leqslant \rho^2$, where $\varepsilon$ is chosen so small that the segment $\xi_2 = \xi_3 = 0$ intersects the surface $h = \varepsilon$. The bifurcation diagram consists of a disk $\xi_1 = 0$ and a segment $\xi_2 = \xi_3 = 0$; it is presented in Figure 5.

**3. cs-point.** Here one has

$$H = h(\xi_1, \xi_2, \xi_3) = \omega\xi_1 + \lambda_2\xi_2 + \lambda_3\xi_3 + \cdots.$$

Since $\xi_1 \geqslant 0$, for $\sigma$ we take the region $0 \leqslant \xi_1 \leqslant d$, $\xi_2^2 + \xi_3^2 \leqslant \rho^2$. Here there are three four-dimensional invariant symplectic disks filled with two-dimensional orbits:

$$\Sigma_4^1 : x_1 = y_1 = 0, \qquad \Sigma_4^2 : x_2 = y_2 = 0, \qquad \Sigma_4^3 : x_3 = y_3 = 0,$$

and three two-dimensional symplectic disks filled with one-dimensional orbits:

$$\Sigma_2^3 : x_1 = y_1 = x_2 = y_2 = 0, \qquad \Sigma_2^2 : x_1 = y_1 = x_3 = y_3 = 0,$$
$$\Sigma_2^1 : x_2 = y_2 = x_3 = y_3 = 0.$$

Any two-dimensional disk is the transverse intersection of two four-dimensional ones. Their images under the modified momentum mapping are two rectangles $\xi_2 = 0$ and $\xi_3 = 0$, and the disk $\xi_1 = 0$, respectively (for four-dimensional manifolds), and the segments $\xi_1 = \xi_2 = 0$, $\xi_1 = \xi_3 = 0$, $\xi_2 = \xi_3 = 0$, respectively. The bifurcation diagram is shown in Figure 5.

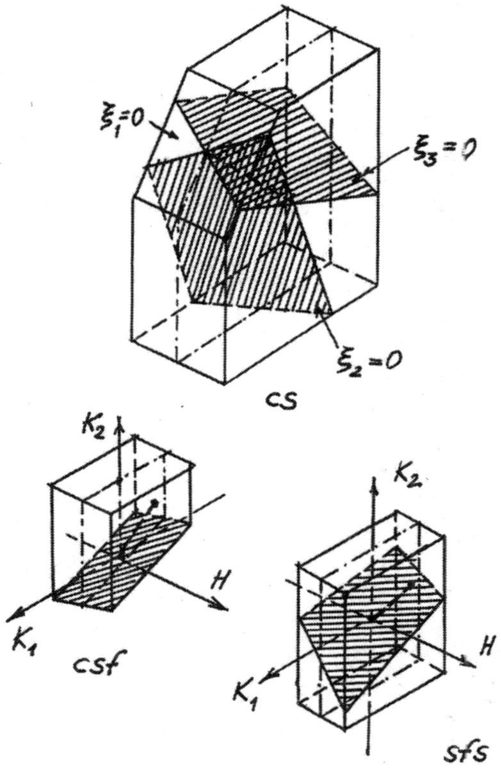


FIGURE 5

**4. sfs-point.** Here

$$H = h(\xi_1, \xi_2, \xi_3) = \alpha\xi_1 + \beta\xi_2 + \lambda\xi_3 + \cdots .$$

Near such a point, only two local invariant symplectic submanifolds exist, a four-dimensional one $\Sigma_4$: $x_3 = y_3 = 0$ filled with two-dimensional action orbits, and a two-dimensional one, $\Sigma_2$: $x_1 = y_1 = x_2 = y_2 = 0$, filled with one-dimensional orbits. Thus, in the space of variables $\xi_1$, $\xi_2$, $\xi_3$ the bifurcation diagram consists of a disk $\xi_3 = 0$ and a segment $\xi_1 = \xi_2 = 0$. Since $\alpha\beta\lambda \neq 0$ the equation $h = 0$ determines a 2-disk transversal to the segment $\xi_1 = \xi_2 = 0$ and to the disk $\xi_3 = 0$. For $\sigma$ we take the region $|h| \leqslant \varepsilon$, $f_1^2 + f_2^2 \leqslant \rho^2$, where $\varepsilon$ is chosen in such a way that the traces of the segment $\xi_2 = \xi_3 = 0$ belong to the inner part of the surfaces $|h| = \pm\varepsilon$ and the disk $h = 0$ does not cross both disks $\xi_3 = \pm d$. The related diagram is depicted in Figure 5.

**5. s-point.** Here

$$H = h(\xi_1, \xi_2, \xi_3) = \lambda_1\xi_1 + \lambda_2\xi_2 + \lambda_3\xi_3 + \cdots .$$

There are three local invariant symplectic four-dimensional submanifolds:

(1) $x_1 = y_1 = 0$, (2) $x_2 = y_2 = 0$, (3) $x_1 = y_1 = x_2 = y_2 = 0$,

and three local symplectic four-dimensional submanifolds:

$$(1)\quad x_1 = y_1 = x_2 = y_2 = 0, \qquad (2)\quad x_1 = y_1 = x_3 = y_3 = 0,$$
$$(3)\quad x_2 = y_2 = x_3 = y_3 = 0.$$

Any two four-dimensional manifolds intersect along one of the two-dimensional ones. In the space $\xi_1$, $\xi_2$, $\xi_3$, this gives as the bifurcation diagram three 2-disks $\xi_1 = 0$, $\xi_2 = 0$, $\xi_3 = 0$, and the segments which are lines of their intersection. For $\sigma$, one can choose the region $|\xi_i| \leqslant d_i$, $i = 1, 2, 3$.

## 4. Orbit structure in an extended neighborhood of an eh-point

To be definite, we assume that $\omega_1 > 0$. Near the point $p$ (in $U$) only four one-dimensional orbits, (forming a "cross" $x_1 = y_1 = x_2 = y_2 = x_3y_3 = 0$) contain $p$ in their closure. These points exhaust the set of solutions of the system $H = 0$, $F_1 = 0$, $F_2 = 0$. These four orbits belong to a two-dimensional local symplectic manifold $\Sigma_2^{(3)}$: $x_1 = y_1 = x_2 = y_2 = 0$, and they are the separatrices of a saddle singular point $p$ for the restriction of the vector field $X_H$ to this submanifold. Let us continue these trajectories along the $X_H$-flow. The limit sets of these trajectories are some singular points (see the assumptions for the Poisson action), so that, due to the uniqueness of the singular point $p$ in the connected part of the level $H = 0$, we conclude that each separatrix comes back to $U$ and coincides with one of these separatrices, forming a loop. It is natural that for $X_H$ one unstable separatrix merges with one of two stable ones. Hence, one gets two separatrix loops. These two loops ("figure eight" $E$) exhaust the entire set of solutions for the system $H = 0$, $F_1 = 0$, $F_2 = 0$ inside the extended neighborhood $V$, while $V$ itself is a small enough neighborhood of $E$.

It is not too hard to verify, as in [**8**], that the one-dimensional orbits (from $\Sigma_2^{(3)}$) passing near the separatrices return to $\Sigma_2^{(3)}$, forming closed orbits. Therefore, the two loops obtained belong to a smooth two-dimensional symplectic manifold diffeomorphic to a disk with two disjoint inner disks removed. We still denote it by $\Sigma_2^{(3)}$.

To study the structure of the IVF inside an extended neighborhood $V$ of $p$, it is convenient to divide $V$ into two parts, local and global ones. To do this, we take four local cross-sections to $X_H$-flow near the local one-dimensional separatrices. In the Vey–Eliasson coordinates, for the sections one can choose the five-dimensional disks $N_\pm^s$: $x_3 = \pm d$ (for the stable separatrices) and $N_\pm^u$: $y_3 = \pm d$ (for the unstable ones). Each such disk carries a foliation formed by the traces on this disk of the action orbits, which are transverse to this disk. This foliation is symplectically diffeomorphic to a Liouville foliation generated by the reduced Hamilton functions $\bar f_1$, $\bar f_2$ in the reduced phase space (the reduction is made by means of the Hamiltonian $H$ since it does not degenerate near sections). Easy calculations show that $\bar f_1$, $\bar f_2$ take the following form in the reduced variables:

$$\bar f_1 = \frac{\mu}{\lambda} c + O(c^2) + \frac{\nu_1\lambda - \mu\omega_1}{\lambda}\,\xi_1 + \frac{\nu_2\lambda - \mu\omega_2}{\lambda}\,\xi_2 + \cdots,$$
$$\bar f_2 = \frac{\kappa}{\lambda} c + O(c^2) + \frac{\rho_1\lambda - \kappa\omega_1}{\lambda}\,\xi_1 + \frac{\rho_2\lambda - \kappa\omega_2}{\lambda}\,\xi_2 + \cdots,$$

with $\xi_i = (x_i^2 + y_i^2)/2$, as above. Since $\lambda \neq 0$ and

$$\Delta = \begin{vmatrix} \omega_1 & \omega_2 & \lambda \\ \nu_1 & \nu_2 & \mu \\ \rho_1 & \rho_2 & \kappa \end{vmatrix} \neq 0,$$

it follows that the determinant $\Delta_1$ does not vanish,

$$\Delta_1 = \begin{vmatrix} \dfrac{\nu_1\lambda - \mu\omega_1}{\lambda} & \dfrac{\nu_2\lambda - \mu\omega_2}{\lambda} \\ \dfrac{\rho_1\lambda - \kappa\omega_1}{\lambda} & \dfrac{\rho_2\lambda - \kappa\omega_2}{\lambda} \end{vmatrix} = \frac{\Delta}{\lambda} \neq 0.$$

Hence, the singular point of the reduced action is a simple and elliptic one. One can choose its neighborhood as in [8] consistently for all $|c|$ small enough. Actually this gives another way of constructing an extended neighborhood of $p$, by taking the preimage of such a neighborhood under the momentum mapping.

In particular, all reduced action orbits except for those belonging to the 2-disks given in reduced variables as $x_1 = y_1 = 0$ or $x_2 = y_2 = 0$, are 2-tori. On the sections they correspond to traces of Lagrangian 3-tori if $y_3 \neq 0$ (on $N_\pm^s$) or $x_3 \neq 0$ (on $N_\pm^u$), or to traces on $N_\pm^s$ of strong stable manifolds of saddle 2-tori lying on the center manifold $W^c = \Sigma_4^{(3)}$ if $y_3 = 0$, or to traces on $N_\pm^u$ of strong unstable manifolds of the same 2-tori on this center manifold if $x_3 = 0$. Closed one-dimensional orbits lie on 2-disks (they correspond to traces of 2-tori of the IVF if $y_3 \neq 0$ (on $N_\pm^s$) or $x_3 \neq 0$ (on $N_\pm^u$)). These 2-tori belong, as we shall see, to invariant four-dimensional symplectic submanifolds obtained by continuing $\Sigma_4^{(1)}$, $\Sigma_4^{(2)}$ along the $X_H$-flow.

Our next goal is to extend, using the $X_H$-flow, the orbits whose traces on $N_\pm^s$, $N_\pm^u$ are closed curves. They correspond to strong stable and strong unstable manifolds of closed trajectories on the local invariant symplectic disks $\Sigma_2^{(1)}$ and $\Sigma_2^{(2)}$ (see Section 3). In $U$, the vector field $X_H$ has a four-dimensional center manifold $W^c$. This manifold is hyperbolic in the sense that $W^c$ is the transverse intersection of two 5-disks, $W^{cs}$ ($y_3 = 0$), and $W^{cu}$ ($x_3 = 0$), such that the trajectories in $W^{cs}$ exponentially tend to $W^c$ as $t \to \infty$, and in $W^{cu}$ trajectories tend to $W^c$ as $t \to -\infty$. The restriction of $X_H$ to $W^c$ gives a 2D IVF, $p$ being a simple singular point of elliptic type for the related Poisson action, since one of the two determinants $\omega_1\nu_2 - \omega_2\nu_1$ or $\omega_1\rho_2 - \omega_2\rho_1$ does not vanish because of the simplicity of the point $p$ in the initial IVF. In $W^c$ there are also two smooth symplectic 2-disks, $\Sigma_2^{(1)}$, and $\Sigma_2^{(2)}$, foliated into closed orbits. The other orbits of the action of $\Phi$ on $W^c$ are two-dimensional tori. Suppose, to be definite, that $\omega_1\nu_2 - \omega_2\nu_1 \neq 0$. Then $F_1$ is an additional integral for $X_H$ on $W^c$. These two functions $H$, $F_1$ separate tori on $W^c$, i.e., the system of equations $H = c$, $F_1 = f$ selects only one torus. Equations $h(\xi_1, \xi_2, 0) = c$, $f_1(\xi_1, \xi_2, 0) = f$ can be solved with respect to $\xi_1$, $\xi_2$, giving the solutions $\xi_1(c, f)$, $\xi_2(c, f)$. For $|\xi_i|$ small enough (and this is where we work) equations $h(\xi_1, \xi_2, \xi_3) = c^0$, $f_i(\xi_1, \xi_2, \xi_3) = f_i^0$, $i = 1, 2$, are uniquely solved with respect to $\xi_i$, $i = 1, 2, 3$, giving solutions $\xi_i(c^0, f_1^0, f_2^0)$, $i = 1, 2, 3$. This implies that the equations $H = c$, $F_1 = f$, $F_2 = f_2(\xi_1(c, f), \xi_2(c, f), 0)$ have as solutions the points lying on locally stable and locally unstable manifolds of the torus $T_{c,f}^2$ on $W^c$ determined by the equations $\xi_1 = \xi_1(c, f)$, $\xi_2 = \xi_2(c, f)$. In $U$ these points belong to the set determined by the relations $x_3 y_3 = 0$, $x_1^2 + y_1^2 = 2\xi_1(c, f)$, $x_2^2 + y_2^2 = 2\xi_2(c, f)$, and this set consists of two local smooth three-dimensional manifolds diffeomorphic to $T^2 \times I$, one of them given by $x_3 = 0$ ($W^u(T_{c,f}^2)$), and the other one by $y_3 = 0$

$(W^s(T^2_{c,f}))$. The strong stable three-dimensional submanifold $W^s(T^2_{c,f})$ intersects the sections $N^s_\pm$, giving two traces, the 2-tori $T^s_\pm$. The same is true for $W^u(T^2_{c,f})$ and $N^u_\pm$. Every homoclinic loop in $E$ defines a correspondence between one of the $N^u_\pm$ and one of the $N^s_\pm$, so that near two loops (and hence, inside the extended neighborhood) two global Poincaré maps are defined and the torus which is the trace of $W^u(T^2_{c,f})$ on $N^u_+$ is transformed by the global Poincaré map onto the unique torus on $N^s_+$ (or $N^s_-$) which is the trace of $W^s(T^2_{c,f})$ on this section. The same is true for the related traces for the other loop.

In $U$ there are also invariant four-dimensional symplectic disks $\Sigma_4^{(1)}$ and $\Sigma_4^{(2)}$. The intersections $W^c \cap \Sigma_4^{(1)} = \Sigma_2^{(2)}$ and $W^c \cap \Sigma_4^{(2)} = \Sigma_2^{(1)}$ are filled with periodic trajectories of $X_H$ (and one-dimensional orbits of $\Phi$). Let us choose a trajectory $\gamma$ on $\Sigma_2^{(1)}$. For this trajectory we have $\xi_1 = \xi_1^0 > 0$, $\xi_2 = 0$, $\xi_3 = 0$, and, hence, $h(\xi_1^0, 0, 0) = c^0$, $f_i(\xi_1^0, 0, 0) = f_i^0$, $i = 1, 2$, so that $\xi_1(c^0, f_1^0, f_2^0) = \xi_1^0$, $\xi_2(c^0, f_1^0, f_2^0) = 0$, $\xi_3(c^0, f_1^0, f_2^0) = 0$. Thus, in $U$ the orbits that contain $\gamma$ in their closure are two cylinders $x_1^2 + y_1^2 = 2\xi_1^0$, $x_2 = y_2 = x_3 = 0$ or $x_1^2 + y_1^2 = 2\xi_1^0$, $x_2 = y_2 = y_3 = 0$. Now let us set $\xi_3 = 0$ and $\xi_2 = \xi_2^0 \neq 0$, the latter being small enough, and $\xi_1 = \xi_1^0$. Then $h(\xi_1^0, \xi_2^0, 0) = c(\xi_2^0)$, $f_1(\xi_1^0, \xi_2^0, 0) = f_1(\xi_2^0)$, and the equations $h(\xi_1, \xi_2, \xi_3) = c(\xi_2^0)$, $f_i(\xi_1, \xi_2, \xi_3) = f_i(\xi_2^0)$, $f_2(\xi_2^0) = f_2(\xi_1^0, \xi_2^0, 0)$ determine a unique torus on $N^s_+$ and $N^u_+$, respectively, these tori being traces of stable and unstable manifolds of the torus $T^2_{c,f} \subset W^c$, $c = c(\xi_2^0)$, $f = f_1(\xi_2^0)$ (we assume here, for definiteness, that the homoclinic loop we consider joins $N^s_+$ and $N^u_+$). Since $\xi_2^0$ is small, these tori are close to the closed curves $x_1^2 + y_1^2 = 2\xi_1^0$, $x_2^2 + y_2^2 = 2\xi_2^0$, $x_3 = d$, $y_3 = 0$ or $x_3 = 0$, $y_3 = d$. Letting $\xi_2^0$ tend to 0, we conclude that the two-dimensional cylinder $x_1^2 + y_1^2 = 2\xi_1^0$, $x_2 = y_2 = x_3 = 0$ being continued along the $X_H$-trajectories merges with the cylinder $x_1^2 + y_1^2 = 2\xi_1^0$, $x_2 = y_2 = y_3 = 0$.

Similar considerations show that $\Sigma_4^{(1)}$ ($\Sigma_4^{(2)}$, respectively), when continued along the $X_H$-trajectories inside the extended neighborhood merges with itself, forming a smooth symplectic four-dimensional submanifold, which is denoted $\Sigma_4^{(1)}$ ($\Sigma_4^{(2)}$) as well. Thus the following proposition holds.

PROPOSITION 1. *In the extended neighborhood $V$ of any eh-point $p$ there are two transversely intersecting four-dimensional submanifolds $\Sigma_4^{(1)}$, $\Sigma_4^{(2)}$ such that the restriction of $X_H$ to $\Sigma_4^{(i)}$, $i = 1, 2$, gives an IVF with the saddle-center point at $p$, each $\Sigma_4^{(i)}$ being an extended neighborhood of this point. For the initial IVF, the local center-stable submanifold $W^{cs}$ and the center-unstable submanifold $W^{cu}$ merge upon continuation along the $X_H$-flow, forming a set diffeomorphic to $D^4 \times E$, where $E$ is the "figure eight" curve.*

In fact, this proposition describes all the orbits that belong to the preimage of the bifurcation diagram $B$ with respect to the momentum mapping. It is clear that the remaining orbits in $V$ are three-dimensional tori, and their number on the joint level of three integrals $H$, $F_1$, $F_2$ are determined in accordance with Figure 4. If one fixes a point $a$ in $\mu(V) \setminus B$, then $\mu^{-1}(a)$ is the disjoint union of one or two 3-tori. If one moves along a path in $\mu(V)$, then nothing changes until the curve crosses the bifurcation diagram. Due to the results in [**26**], the following bifurcations occur when crossing $B$ in Figure 4.

1. If a path starts at a point $a \in B$ that corresponds to $\mu(\Sigma_4^{(1)} \setminus \Sigma_4^{(3)})$ or $\mu(\Sigma_4^{(2)} \setminus \Sigma_4^{(3)})$, then $\mu^{-1}(a)$ is a 2-torus (or two such tori), and the preimage of

this path is diffeomorphic to $D^2 \times T^2$ with its foliation generated by the standard foliation of $D^2$ of the type occurring near a center point.

2. Along a path that crosses $B$ at a point on $\mu(\Sigma_4^{(3)})$, a bifurcation transforming two 3-tori into one 3-torus, or conversely, occurs. This bifurcation can be described by means of the foliation in $T^2 \times \Sigma_2^{(3)}$ generated by the foliation in $\Sigma_2^{(3)}$ into one-dimensional orbits described above.

When studying the extended neighborhood of an eh-point, we shall distinguish two different cases: (i) $\omega_1\omega_2 > 0$ and (ii) $\omega_1\omega_2 < 0$.

(i) This subcase is similar to that of two degrees of freedom in the sense that the structure of the levels $H = c$, $|c| \leqslant c^*$, can be described as in [**8**]. Recall that we always may assume $\lambda > 0$ and set, to be definite, $\omega_1 > 0$, $\omega_2 > 0$. Using the local representation (1) and the Morse lemma, we obtain

PROPOSITION 2. *There is a neighborhood of $p$ such that any level $H = c$, $c < 0$, consists of two components, each of which is homeomorphic to $D^4 \times I$. The level $H = 0$ is a wedge*[1] *of two $D^4 \times I$ glued at an inner point. A level $H = c$, $c > 0$, is the connected sum of two $D^4 \times I$.*

It follows from this proposition that the intersection of an extended neighborhood with $H = c$, $c < 0$, is either a disconnected set, in which case it is the disjoint union of two $D^4 \times S^1$ (case 1), or a connected set, in which case it is $D^4 \times S^1$ (case 2). When $H = 0$, both manifolds $D^4 \times S^1$ (case 1) are glued at two inner points of each copy, in case 2 two different inner points in $D^4 \times S^1$ are identified. Finally, if $H = c$, $c > 0$, the manifold $V_c = V \cap \{H = c\}$ is the connected sum of two $D^4 \times S^1$ (case 1), and the connected sum of $D^4 \times S^1$ and $S^4 \times S^1$ (case 2).

Another way to see the existence of two isoenergetically distinct cases is to look at the restriction of the IVF to the submanifolds $\Sigma_4^{(1)}$ and $\Sigma_4^{(2)}$ on which the restricted IVF has a saddle-center singular point. Their bifurcation diagrams correspond to faces of the whole diagram ($\xi_1 = 0$ or $\xi_2 = 0$). It is clear from Figure 4 and results of [**8**] that in this case we have the same isoenergetical structures for both submanifolds. So two possibilities can occur, since there are two isoenergetical structures for a saddle-center point.

(ii) In this subcase the restrictions of the IVF to $\Sigma_4^{(1)}$ and $\Sigma_4^{(2)}$ give IVFs with different isoenergetical structures (see Figure 4). Having in mind that the change $H \to -H$ (or $t \to -t$) leads to isoenergetically equivalent systems, we come to the conclusion that only one isoenergetical equivalence class exists in this case. Here it is also possible to describe topological (and even differential) types of level sets $H = c$, but I do not know any standard models for these manifolds, only a construction (a specific gluing of two standard blocks).

Summarizing one can state

THEOREM 3. *Let a 3D IVF have a simple eh-point. Then there are three and only three isoenergetically different such IVFs. These cases are distinguished by the sign $\omega_1\omega_2$ in the local representation of $H$ and, if $\omega_1\omega_2 > 0$, by the number of periodic trajectories* (1 *or* 2) *on $\Sigma_2^{(3)}$ at the level $H = c$, where* $\operatorname{sign}(c) = \operatorname{sign}(\omega_1)$.

---

[1]By a wedge of two topological spaces $(X_i, p_i)$, $i = 1, 2$, with distinguished points $p_i$, one means a topological space $X_1 \cup X_2$ with the identification $p_1 = p_2$ and the quotient topology [**27**].

## 5. Orbit structure in an extended neighborhood of a csf-point

In this case only two local two-dimensional orbits, $S$ and $W$, containing $p$ in their closure exist. For $X_H$, these orbits are locally stable and unstable two-dimensional manifolds for the restriction of $X_H$ to the unique local four-dimensional symplectic disk $\Sigma_4$ (see Section 3). For this restriction $p$ is a saddle-focus singular point and is of focus-focus type for the corresponding Poisson action of $\mathbb{R}^2$ [**7**]. The limit set for a two-dimensional orbit can be either a one-dimensional orbit or a singular point, due to the assumption on decreasing dimension (see Section 2). If an IVF is of "generic type" (see the related discussion in [**17**]), then the limit set must be a singular point (otherwise, one can change the IVF in such a way that the new IVF will have only a singular point as the limit set), and due to the uniqueness of the singular point $p$ in the connected part of the set determined by equations $H = F_1 = F_2 = 0$, this point is $p$. But locally near $p$ this set consists only of the two aforementioned orbits completed by the point $p$, i.e., two transverse disks in $\Sigma_4$ ("two-dimensional cross"). This implies that the continuation of one local orbit with respect to the action will merge with the other one.

In $\Sigma_4$ let us select two local three-dimensional cross-sections $N_S$, $N_W$ to $S$, $W$, respectively. This can be done as in [**12**] for the restriction of $X_H$ to $\Sigma_4$. Then any 2-orbit close to $S$ intersects $N_S$ and, since the part of $S$ between $N_S$ and $N_W$ is a compact annulus, it follows that this is also true for any nearby 2-orbit. This consideration implies that nearby two-dimensional orbits from $\Sigma_4$ can be continued along the compact set between $N_S$ and $N_W$. Thus, the continuation of $\Sigma_4$ forms a four-dimensional compact symplectic manifold with boundary that is nothing but the extended neighborhood of a focus-focus singular point as in [**8**].

Now choose five-dimensional cross-sections to the trajectories of the vector field $X_H$ (and hence, to the action orbits as well) passing close to the disks $S$ and $W$. To that end, we first choose a circle in $S$, namely $l$: $x_1 = y_1 = y_2 = y_3 = 0$, $x_2^2 + x_3^2 = \rho_s^2$, with $\rho_s$ small enough. This circle is transverse to trajectories on the disk $S$. Therefore, the submanifold $N^s$: $x_2^2 + x_3^2 = \rho_s^2$, $x_1^2 + y_1^2 + y_2^2 + y_3^2 < \delta_s^2$ is transverse to all trajectories that are sufficiently close to $S$ (not only for those in $\Sigma_4$). Traces of action orbits on $N^s$ generate a foliation whose structure can be described as follows. Since $S$ is two-dimensional, there are two independent integrals on $S$. Since $p$ is simple and $\omega\alpha\beta \neq 0$, one can always assume that the independent integrals are $H$, $F_1$ (changing the numbering, if necessary). Thus, on the joint level $H = c$, $F_1 = f$, being a smooth 4-disk locally, near a point $m \in l$, one can perform the reduction procedure using the integrals $H$, $F_1$. Then one obtains a two-dimensional symplectic disk with the 1D IVF generated by the reduced third integral $\hat{f}_2$. This IVF has a center singular point at the point corresponding to $m$, the other orbits being closed. The reduction smoothly depends on two parameters $c$, $f$; so, returning to the section $N^s$ on the level $H = c$, $F_1 = f$, we obtain a standard foliation of solid torus $D^2 \times S^1$ into 2-tori shrinking to a unique circle which is the trace of a two-dimensional action orbit from $\Sigma_4$. This whole picture depends smoothly on $c$, $f$, so this gives a foliation of the whole set $N^s$.

The same considerations are valid for $W$ and the related section $N^u$. Any trajectory of $X_H$ beginning at a point on $N^u$ reaches $N^s$ in a finite time, uniformly bounded for all such points. Thus, the Poincaré map transforms the 2-torus $H = c$, $F_1 = f$, $F_2 = q$ into a 2-torus in $N^s$ given by the same relations. But the set of these constants determines a unique torus. In addition, let us describe possible twistings that occur while passing from the torus in $N^u$ to the one in $N^s$. The solid torus $H = c$, $F_1 = f$ on $N^u$ is transformed to the solid torus in $N^s$ in such a way that its central circle (the trace of the singular leaf) is transformed to the central circle and any disk transverse to the central circle is transformed to a disk transverse to the similar circle in the solid torus in $N_S$. This means that any torus can be twisted under this transformation only along one basic circle nonhomotopic to the central circle in the solid torus. This leads to the following assertion.

THEOREM 4. *All IVFs with cfs-points are isoenergetically equivalent in some of their extended neighborhoods.*

COROLLARY. *The Hamiltonian of the direct product of two IVFs, $H = H_1+H_2$, with*

$$H_1 = \frac{\omega(x_1^2 + y_1^2)}{2}, \qquad H_2 = \frac{p_2^2 + p_3^2}{2} + p_3q_2 - p_2q_3 - \frac{q_2^2 + q_3^2}{2} + \frac{(q_2^2 + q_3^2)^2}{4}$$

*is the standard model Hamiltonian with this behavior.*

## 6. Orbit structure in an extended neighborhood of a cs-point

In the local representation $h = \omega\xi_1 + \lambda\xi_2 + \gamma\xi_3 + \dots$, $\omega\lambda\gamma \neq 0$, without loss of generality one may assume $\lambda$, $\gamma$ to be positive; this can always be achieved by a local symplectic transformation. We also assume, to be definite, that $\omega > 0$. As explained in Section 3, locally near $p$ there are three four-dimensional invariant symplectic disks,

$$\Sigma_4^{(1)}: \quad x_1 = y_1 = 0, \qquad \Sigma_4^{(2)}: \quad x_2 = y_2 = 0, \qquad \Sigma_4^{(3)}: \quad x_3 = y_3 = 0.$$

The restrictions of $X_H$ to the two latter disks give saddle-center points at $p$, and a saddle for the former (see [**12, 13**]). There are also three two-dimensional invariant symplectic disks,

$$\Sigma_2^{(1)}: \quad x_2 = y_2 = x_3 = y_3 = 0, \qquad \Sigma_2^{(2)}: \quad x_1 = y_1 = x_3 = y_3 = 0,$$
$$\Sigma_2^{(3)}: \quad x_1 = y_1 = x_2 = y_2 = 0$$

with saddles at $p$ for the restrictions of $X_H$ to the two latter disks and a center for the former. As above, we can extend the one-dimensional separatrices from $\Sigma_2^{(2)}$ and $\Sigma_2^{(3)}$ to obtain four homoclinic loops. Using considerations similar to [**8**], one can prove the following

LEMMA. *Separatrices from $\Sigma_2^{(i)}$ can return only to the same $\Sigma_2^{(i)}$, $i = 2, 3$.*

This implies that the neighborhoods of these loops on both $\Sigma_2^{(i)}$, $i = 2, 3$, are filled with closed orbits and form two thick "figure eight" sets that will be denoted by the same letters as the related disks. Since $H(p) = 0$, the restriction of $H$ to $\Sigma_2^{(i)}$, $i = 2, 3$, gives a function that vanishes at the loops, so that the "figure eight" divides the corresponding $\Sigma_2^{(i)}$ into two parts; one of them is an annulus (the closed orbit lying on the annulus boundary will be called the outer boundary of $\Sigma_2^i$), and the other consists of two connected components (two annuli). The Hamiltonian $H$ has a definite sign on the outer boundary.

Now we continue the orbits from $U$ near the global pieces of $\Sigma_2^{(i)}$, $i = 2, 3$. This continuation is made by means of the $X_H$-flow. Let us fix, for instance, $\Sigma_2^{(2)}$ and a loop $\gamma$ on it. Then one can choose two five-dimensional local cross-sections

$N^s$, $N^u$ to the ingoing and outgoing pieces of the loop with respect to $X_H$. The sections $N^s$, $N^u$ are foliated by levels $H = c$ into four-dimensional symplectic disks $N_c^s$, $N_c^u$. By the transversality of the flow trajectories to $N_c^s$, $N_c^u$, and by the invariance of $H = c$, all action orbits close to $\gamma$ are transverse to $N_c^s$, $N_c^u$ and their traces on $N_c^s$, $N_c^u$ generate a foliation. This foliation on the symplectic disk $N_c^s$ ($N_c^u$) is symplectomorphic to the Liouville foliation given by the reduced action of $\mathbb{R}^2$ generated by the reduced Hamiltonians $\hat{f}_1$, $\hat{f}_2$, where the reduction is carried out by means of the Hamiltonian $H$. As above, it can be easily shown that the reduced action (depending on the parameter $c$) has a saddle-center singular point at $\gamma \cap N_c^s$ ($\gamma \cap N_c^u$, respectively), and the same is true for all points that are traces of one-dimensional orbits from $\Sigma_2^{(2)}$ on $N_c^s$ (the parameter $c$ enumerates these points). The $X_H$-flow transforms the foliation on $N_c^u$ to the one on $N_c^s$. Let us recall [**12**] that near a saddle-center there are two two-dimensional invariant symplectic disks filled with one-dimensional orbits, one of them carries a center point foliation, the other one carries a saddle point foliation. It is evident that the flow transforms the center disk in $N_c^u$ to that in $N_c^s$, and the same is true for saddle disks. If we take center disks for all $c$ small enough, we obtain upon continuing an extended neighborhood of a saddle-center point in $\Sigma_4^{(2)}$ (the neighborhood contains $\Sigma_2^{(1)}$ and $\Sigma_2^{(3)}$). Thus, globally, two such four-dimensional invariant symplectic submanifolds, being extended neighborhoods of saddle-center points, intersect each other transversely along a two-dimensional symplectic disk $\Sigma_2^{(1)}$.

In the same way, it can be shown that the four-dimensional invariant symplectic disk $\Sigma_4^{(1)}$, when continued, forms a global four-dimensional symplectic submanifold, denoted again by $\Sigma_4^{(1)}$. being an extended neighborhood of a saddle-saddle point on $\Sigma_4^{(1)}$. The set $\Sigma_4^{(1)}$ transversely intersects $\Sigma_4^{(2)}$ along $\Sigma_2^{(3)}$ and $\Sigma_4^{(3)}$ along $\Sigma_2^{(2)}$. These three four-dimensional symplectic submanifolds in the extended neighborhood $V$ constitute the preimage of the bifurcation diagram $B$ in $V$. The other orbits in $V$ are Liouville 3-tori.

As is known (see [**12**]), there are 8 types of isoenergetically nonequivalent IVFs for a saddle-saddle point. They are distinguished by the orientability types of the four loops and, for a fixed collection of loops, by the signs of the function $H$ on the outer boundary of the two thick "figure eights". Thus the restriction of $X_H$ on $\Sigma_4^{(1)}$ gives an IVF of one of 8 possible types. After that the signs of $H$ on $\Sigma_2^{(2)}$ and $\Sigma_2^{(3)}$ are uniquely determined (see Figure 5). Recall that by our assumption $\omega > 0$, the Lyapunov periodic trajectories in $\Sigma_2^{(3)}$ lie in the levels $H > 0$. Then, by the results of [**12**], the isoenergetical type of the restriction of $X_H$ to $\Sigma_4^{(2)}$ is completely determined. The same is true for $\Sigma_4^{(1)}$.

As is easily seen, every two-dimensional torus in $\Sigma_4^{(1)}$ is elliptic with respect to the directions transverse to $\Sigma_4^{(1)}$ (in the corresponding four-dimensional joint level of the two independent integrals containing this torus, the torus is indeed elliptic). But the 2-tori in $\Sigma_4^{(2)}$ and $\Sigma_4^{(3)}$ are hyperbolic with respect to transverse directions. Therefore, they can be orientable or nonorientable, and this is determined by the orientability types of the corresponding loops in $\Sigma_4^{(1)} \cap \Sigma_2^{(i)}$, $i = 2, 3$.

In the set $\mu(V)$, the part of $B$ corresponding to $\xi_1 = 0$ (i.e., $\mu(\Sigma_4^{(1)}) = B_1$) is related to case 1 in Section 4, and $B_i = \mu(\Sigma_4^{(i)})$, $i = 2, 3$, is related to the case 2.

Thus, the preimage of a thin slice near $B_1$ is diffeomorphic to $\Sigma_4^{(1)} \times D^2$, where $D^2$ is taken with a foliation near the center point. Therefore, the orientability type of the saddle tori in $\Sigma_4^{(i)}$, $i = 2, 3$, completely depends on the orientability type of the loops in $\Sigma_4^{(1)}$ from which these tori appear. Another thing is the number of Liouville 3-tori, which constitute the level $H = c$, $F_1 = f_1$, $F_2 = f_2$. It depends on which component of $\mu(V) \setminus B$ the point $(c, f_1, f_2)$ lies. There are 4 such components, and each of them contains in its closure only one of the 4 components of $B_1 \setminus (B_2 \cup B_3)$. This implies that the number of Liouville tori is completely determined by the number of 2-tori in the corresponding component of $B_1 \setminus (B_2 \cup B_3)$. From these considerations the following theorem can be derived.

THEOREM 5. *There are 8 types of different isoenergetic equivalence classes in an extended neighborhood of any cs-point. Each class can be realized by the direct product Hamiltonian $H = H_1 + H_2$, where $H_1 = (x_1^2 + y_1^2)/2$, and $H_2$ realizes one of the 8 types of isoenergetic behavior in the extended neighborhood of a saddle-saddle singular point.*

## 7. Orbit structure in an extended neighborhood of an sfs-point

One can always assume that $\alpha$, $\lambda$ in the local representation of

$$H = h = \alpha\xi_1 + \beta\xi_2 + \lambda\xi_3 + \cdots$$

are positive. A singular leaf $H = F_1 = F_2 = 0$ locally consists of four three-dimensional Lagrangian disks $W^s$, $W_1^c$, $W_2^c$, $W^u$ given by the equations

$$y_1 = y_2 = y_3 = 0 \quad (W^s), \qquad x_1 = x_2 = y_3 = 0 \quad (W_1^c),$$
$$y_1 = y_2 = x_3 = 0 \quad (W_2^c), \qquad x_1 = x_2 = x_3 = 0 \quad (W^u).$$

The first and fourth are respectively stable and unstable local manifolds of the singular point $p$ for $X_H$, and $p$ is a saddle singular point for the restrictions of $X_H$ to the two others disks. For the sequel, we need to describe the local structure of the singular leaf near a point $m$ lying on one-dimensional orbits. One may use the usual reduction procedure [**20**], but it is more convenient in this case to utilize the Vey–Eliasson coordinates. To be definite, assume that $m$ has the following coordinate representation $x_1 = x_2 = y_1 = y_2 = y_3 = 0$, $x_3 = x_3^0 \neq 0$. From the equation $h(\xi_1, \xi_2, \xi_3) = 0$ one derives

$$\xi_3 = -\frac{\alpha}{\lambda}\xi_1 - \frac{\beta}{\lambda}\xi_2 - O_2(\xi_1, \xi_2).$$

Substituting $\xi_3$ into $F_1$ and $F_2$, one obtains a two-degrees-of-freedom system with respect to $(x_1, x_2, y_1, y_2)$, depending on $\xi_1$, $\xi_2$ only. The commuting integrals are reduced functions $\hat{f}_1$, $\hat{f}_2$. Since these integrals depend on the variables via $\xi_1$, $\xi_2$ and have the form

$$\hat{f}_1 = \frac{\alpha_1\lambda - \alpha\lambda_1}{\gamma}\xi_1 + \frac{\beta_1\lambda - \beta\lambda_1}{\lambda}\xi_2 + O_2(\xi_1, \xi_2),$$
$$\hat{f}_2 = \frac{\alpha_2\lambda - \alpha\lambda_2}{\lambda}\xi_1 + \frac{\alpha_2\lambda - \alpha\lambda_2}{\lambda}\xi_2 + O_2(\xi_1, \xi_2),$$

where the determinant of the coefficients at $\xi_1$, $\xi_2$ is equal to $\Delta/\lambda \neq 0$, it follows that the singular point of the reduced action corresponding to the one-dimensional orbit $l$ through $m$ is a focus-focus, and only two two-dimensional local orbits make up the local singular leaf of this point [**7**]. It is evident that locally near $m$ the structure of the Liouville foliation is the direct product of a reduced local focus-focus foliation and a two-dimensional rectangle, one direction of which relates with the orbit itself (the $x_3$-direction) and the other one with varying $H$. Thus, along $l$, one has two three-dimensional action orbits whose closures intersect each other transversely along $l$ on the level $H = 0$.

Similarly, for a point $m$ on the two-dimensional separatrices

$$W_1: \quad x_3 = y_3 = x_1 = x_2 = 0, \qquad W_2: \quad x_3 = x_3 = y_1 = y_2 = 0$$

there are also two three-dimensional local disks that belong to the level $H = 0$, which also intersect each other along $W_i$, but not transversely. But, in the joint level $H = 0$, $F_1 = 0$ (if $dH$ and $dF_1$ are independent at $m$), or $H = 0$, $F_2 = 0$ (if $dH$ and $dF_2$ are independent at $m$), this intersection is transverse.

Let us now examine the global behavior of the three-dimensional Lagrange disks in order to understand the global structure of the set given by the equations $H = 0$, $F_1 = 0$, $F_2 = 0$. First we extend the separatrices, i.e., the local one-dimensional orbits lying on the "cross" $x_1 = x_2 = y_1 = y_2 = x_3y_3 = 0$. Since locally, near a point on a one-dimensional separatrix, there is only one two-dimensional symplectic submanifold $\Sigma_2$ filled with one-dimensional orbits and the closure of the one-dimensional separatrix contains the only singular point $p$ (it is unique at the level $H = F_1 = F_2 = 0$), it follows that one-dimensional separatrices from $\Sigma_2$, when continued, form two loops ("figure eight"), as in the cases of eh-points and cs-points (see above). It is easily seen that the other one-dimensional orbits from $\Sigma_2$, being close to loops, form closed curves, and thus the continuation of $\Sigma_2$ gives a global two-dimensional symplectic submanifold that is again denoted by $\Sigma_2$. It is diffeomorphic to a thick "figure eight". In the same way, one can extend, by means of the $X_H$-flow, the two-dimensional orbit $W_1$. After extending, we obtain a whole two-dimensional orbit, which is an immersed cylinder $\mathbb{R} \times S^1$ whose closure captures the point $p$ only and contains both local two-dimensional orbits near $p$ (it is homeomorphic to a sphere $S^2$ with two of its points identified).

Next, we extend the other two-dimensional orbits that are close to $W_1$. Simple considerations (related, as in Section 6, to the existence of the local cross-sections $N^s$, $N^u$ to $W^s$ and $W^u$, where all the two-dimensional orbits have their traces, and to the finiteness of the time required by the flow trajectories to pass from $N^u$ to $N^s$) show that the continuation gives a global four-dimensional invariant symplectic submanifold with boundary, still denoted by $\Sigma_4$, filled with two-dimensional action orbits and containing the local 4-disk $x_3 = y_3 = 0$. It is evident that the restriction of the action to $\Sigma_4$ is a Poisson action of the group $\mathbb{R}^2$ with a focus-focus singular point at $p$, and $\Sigma_4$ is an extended neighborhood of this point (see [**12**]). It is clear that $\mu(\Sigma_4)$ coincides with the curvilinear rectangle $R$: $\xi_3 = 0$ of the bifurcation diagram, but $\mu^{-1}(R)$ is not exhausted by $\Sigma_4$ since all orbits in $\Sigma_4$ are saddle orbits, and so, their stable and unstable manifolds also belong to this set.

In order to classify possible types of isoenergetic behavior, let us first determine the types of loops at the level $H = F_1 = F_2 = 0$. To do this, we fix a local invariant Lagrangian disk $W^u$ and extend it by the $X_H$-flow. This disk consists of the point $p$ (zero-dimensional orbit), two local one-dimensional orbits belonging the $y_3$-axis, one two-dimensional orbit (a punctured disk $y_3 = 0$), and two three-dimensional orbits (open semi-disks $y_3 > 0$ and $y_3 < 0$ with the deleted $y_3$-axis). The continuation of a one-dimensional orbit $\gamma$, say, $y_1 = y_2 = 0$, $y_3 > 0$, gives one of two separatrix loops.

Let us extend along $\gamma$ one of the two three-dimensional orbits that contains $\gamma$ in its closure. Suppose it is an orbit $L$ given by the inequality $y_3 > 0$ (a half of a locally unstable manifold of $p$ for $X_H$). Continuing this orbit along a global piece of $\gamma$, we come back into the neighborhood of $p$, where $\gamma$ coincides at the level $H = F_1 = F_2 = 0$ with one of the two $x_3$-semiaxes. This $x_3$-axis is the transverse intersection of two three-dimensional disks, $W_1^c$ and $W^s$, since the chosen semi-axis belongs to the closure in the level $H = 0$ of two three-dimensional orbits. Thus continuation of $L$ will merge locally near $\gamma$ with one of these 3-orbits. It is not hard to show, as in [**8**], that $L$ will merge with the 3-orbit lying in $W_1^c$. It is worth mentioning that $W_1^c$ contains the 2-disk $x_1 = x_2 = x_3 = y_3 = 0$, the same one that divides $W^u$ into the two parts $y_3 > 0$ and $y_3 < 0$.

REMARK 4. It follows from the fact that all short (i.e., homotopic to $\gamma$) periodic trajectories of $X_H$ that are close to $\gamma$ in $\Sigma_2$ are saddle trajectories inside a level of $H$ containing the related orbit, they have stable and unstable manifolds that intersect each other transversally along a periodic orbit in the corresponding level of $H$, and this structure is inherited by the structure at the level $H = F_1 = F_2 = 0$.

Now we can form the following three-dimensional set. It consists of the 2-disk $x_1 = x_2 = x_3 = y_3 = 0$ and the part of $L$ belonging to a neighborhood of $\gamma$. This set is a topological manifold, being a bundle over the circle (i.e., $\gamma$) with fiber $D^2$ (the two-dimensional disk). This bundle can be only of two types, orientable or nonorientable. For instance, for the case of a saddle point of a two-degrees-of-freedom IVF, both such possibilities can occur [**12**]. I believe that the same situation can be realized in this case. Research in this direction is in progress.

**An example.** The direct product of two IVFs gives a structure with orientable loops. Indeed, on the symplectic plane $\mathbb{R}^2$ with coordinates $(x_3, y_3)$ and 2-form $dx_2 \wedge dy_3$, let us consider the Hamiltonian $H_1 = y^2/2 - x^2/2 + x^4/4$ inside the region $B_1$ given by inequality $|H| \leqslant 1/8$. The system has the saddle $(0, 0)$ with two homoclinic loops to it, the other trajectories being closed (a thick "figure eight"). The second IVF is given in the symplectic space $\mathbb{R}^4$ with coordinates $(x_1, x_2, y_1, y_2)$ and 2-form $dx_1 \wedge dy_1 + dx_2 \wedge dy_2$ by the Hamiltonian

$$H_2 = x_2y_2 + x_3y_3 + x_2y_3 - x_3y_2 + [(x_2^2 + x_3^2)^2 + (y_2^2 + y_3^2)^2].$$

This system is integrable (the additional integral is $K = x_2y_3 - x_3y_2$) and generates an action that has a focus-focus at the origin with merged two-dimensional separatrices (see [**14**]). For the region $B_2$, we choose some extended neighborhood of the focus-focus. All orbits in $B_2$ except for the separatrices (and the singular point) are two-dimensional tori. Now in $\mathbb{R}^6$ with the direct product symplectic structure we can take the Hamiltonian $H = H_1 + H_2$ inside the region $B_1 \times B_2$. It is evident that this IVF is the one we need.

Another situation occurs when we extend three-dimensional local disks along $W_1$. Consider the continuation of the three-dimensional disk $x_1 = x_2 = x_3 = 0$ along the two-dimensional orbit $y_3 = 0$. As we already know, this latter orbit lies in $\Sigma_4$ and merges, when continued, with the local two-dimensional orbit $W_2$, forming the whole two-dimensional orbit $W$. Moreover, flow trajectories of $X_H$ pass the global part of $W$ lying outside a small neighborhood of $p$ in finite time. Thus, a part of the orbit close to $W$ can be extended until it reaches a neighborhood of $p$. Two three-dimensional Lagrangian disks $W^s$ and $W_2^c$ intersect along the disk $x_3 = y_1 = y_2 = y_3 = 0$. Hence $W^u$ will merge with one of these disks.

Considerations related to the local structure of the two-dimensional saddle tori filling $\Sigma_4$ allow one to deduce that $W^u$ must merge with $W_2^c$, and similarly, $W_1^c$ must merge with $W^s$. But this merging can be of two types. To understand this, let us observe that $W^u$ and $W_2^c$ both contain a segment of the $y_3$-axis in their closure (these disks intersect locally and merge when globally continued). Thus, taking some orientation of the $y_3$-axis, under global continuation we come to some definite orientation of the same axis. It is clear that we can get the same orientation as well as the opposite one. We call the first case orientable and the other one nonorientable. In the first case every torus $T_0$ from $\Sigma_4$ will be orientable at the joint level $H = c$, $F_i = s$, where $i$ is determined by that function $F_i$ which, together with $H$, is independent of the torus and constants $c$, $s$ are given by $c = H(T_0)$, $s = F_i(T_0)$. Now we will present an example of an IVF with nonorientable saddle 2-tori.

**An example with sfs-point and nonorientable saddle tori.** To construct such an example, we shall follow [**14**], that is, the desired vector field will be obtained by gluing two 6-dimensional submanifolds with boundaries carrying vector fields. Let us consider the two-dimensional Hamiltonian system with the Hamilton function $H_1 = y_3^2/2 - x_3^2/2 + x_3^4/4$ in the region $D_1$: $|H_1| \leqslant 1/8$ with 2-form $\omega_1 = dx_3 \wedge dy_3$, where it has a separatrix "figure eight" (two loops), along with its neighborhood filled with periodic trajectories. As the second Hamilton function, we take the quadratic function $H_2 = x_1y_1 + x_2y_2 + x_1y_2 - x_2y_1 = \xi + \eta$ generating a linear Hamiltonian system with a saddle-focus singular point $O = (0,0,0,0)$. This system is considered in the region $D_2$ given by inequalities $|\xi| \leqslant \delta$, $|\eta| \leqslant 1/2$, $\ln(x_1^2+x_2^2) \leqslant 2$, $\ln(y_1^2+y_2^2) \leqslant 2$ with the standard 2-form $\omega_2 = dx_1 \wedge dy_1 + dx_2 \wedge dy_2$. The region $D_2$ contains two separatrix disks (locally stable and unstable manifolds $W^s$, $W^u$ of $O$) along with a four-dimensional neighborhood of their union. All trajectories of $X_{H_2}$ except those belonging to $W^u$, enter $D_2$ through a cross-section to $W^s$, namely $N^s = \{x_1^2+x_2^2 = e^2\}$. Similarly, all trajectories of $X_{H_2}$, except those belonging $W^s$, leave $D_2$ through a cross-section to $W^u$, namely $N^u = \{y_1^2+y_2^2 = e^2\}$. Inside $D_2$ we distinguish two disjoint subregions, one near $N^s$, denoted $B^+$, and the other one, near $N^u$, denoted $B^-$. Each of them is selected in $D_2$ by the inequalities

$$B^+ = \{e \leqslant x_1^2 + x_2^2 \leqslant e^2,\ y_1^2 + y_2^2 \leqslant Q^2\}, \quad B^- = \{x_1^2 + x_2^2 \leqslant Q^2,\ e \leqslant y_1^2 + y_2^2 \leqslant e^2\},$$

where $Q$ is a constant obeying the inequality $Q^2 \geqslant (4 + \delta^2)e^{-2}$.

Next in the region $D = D_2 \times D_1$ with the 2-form $\omega_2 + \omega_1$, we take the a Hamiltonian $H = H_2 + H_1$, the structure of its flow is clear from the direct product structure. It is an IVF with $F_1 = \xi$, $F_2 = \eta$. The singular point $p$ at the origin is simple since $\Delta = 1$. It should be observed that the vector field $X_{H_1}$ in $D_1$ is symmetric with respect to the symplectic involution $\sigma$: $(x_3, y_3) \to -(x_3, y_3)$.

At the next step of the construction, we identify points of two subregions in $D$ in order to obtain a smooth compact symplectic manifold $M$ with boundary and a complete flow. The boundary of $M$ will be an invariant set of the flow. To make the construction transparent, it is worth recalling that almost all trajectories of the flow of $X_H$ in $D$ enter $D$ through the section $N^s \times D_1$ and leave $D$ through $N^u \times D_1$. Trajectories with a different behavior belong to the sets $W^u \times D_1$ or $W^s \times D_1$. In particular, the set $\{O\} \times D_1$ is invariant with respect to the flow, and it is exactly the set $\Sigma_2$ of one-dimensional orbits of the action.

To construct $M$, we introduce the symplectic coordinates $(R_1, \varphi, \xi, \eta)$ in $B^+$ by setting

$$x_1 = e^{R_1}\cos\varphi, \qquad y_1 = (\xi\cos\varphi - \eta\sin\varphi)e^{-R_1},$$
$$x_2 = e^{R_1}\sin\varphi, \qquad y_2 = (\xi\sin\varphi + \eta\cos\varphi)e^{-R_1}.$$

Then we have

$$\omega_2 = dR_1 \wedge d\xi + d\varphi \wedge d\eta, \qquad H_2 = \xi + \eta, \qquad X_{H_2} = -\frac{\partial}{\partial R_1} - \frac{\partial}{\partial \varphi}.$$

Similarly, in $B^-$ we introduce the coordinates $(\xi, \eta, R_2, \psi)$ as

$$x_1 = (\xi\cos\psi - \eta\sin\psi)e^{-R_2}, \qquad x_2 = (\xi\sin\psi + \eta\cos\psi)e^{-R_2},$$
$$y_1 = e^{R_2}\cos\psi, \qquad y_2 = e^{R_2}\sin\psi,$$
$$\omega_2 = d\xi \wedge dR_2 + d\eta \wedge d\psi, \qquad H_2 = \xi + \eta, \qquad X_{H_2} = \frac{\partial}{\partial R_2} + \frac{\partial}{\partial \psi}.$$

The gluing mapping $\tau$ is now given by

$$\tau\colon \xi \to \xi, \quad \eta \to \eta, \quad R_2 \to \frac{3}{2} - R_1,$$
$$\psi \to 2\pi - \varphi (\operatorname{mod} 2\pi), \quad x_3 \to -x_3, \quad y_3 \to -y_3.$$

This mapping takes $B^- \times D_1$ onto $B^+ \times D_1$, the vector field $X_H$ in $B^- \times D_1$ to $X_H$ in $B^+ \times D_1$. Now we identify every point $m$ in $B^- \times D_1$ with its image $\tau(m)$ in $B^+ \times D_1$. Our formulas imply that we have obtained

(i) a smooth symplectic manifold $M$;
(ii) the function $H$, as well as the functions $\xi$, $\eta$ glued into smooth functions on $M$ that will be denoted with the same letters;
(iii) the IVF $X_H$ generated by $H$ with two additional integrals $\xi$ and $\eta$.

It is easily verified that $M$ realizes the nonorientable case we sought for.

## References


[1] M. Audin, *The topology of torus actions on symplectic manifolds*, Progress in Mathematics, vol. 93, Birkhäuser, Basel, 1991.
[2] A. T. Fomenko, *Integrability and nonintegrability in geometry and mechanics*, Kluwer, Dordrecht, 1988.
[3] L. Bates and R. Cushman, *Global aspects of classical integrable systems*, Birkhäuser, Basel, 1997.
[4] J. Llibre and A. Nunes, *Separatrix surfaces and invariant manifolds of a class of integrable Hamiltonian systems and their perturbations*, Mem. Amer. Math. Soc., vol. 513, Providence, RI, 1994.
[5] L. M. Lerman and Ya. L. Umanskiĭ, *Four-dimensional integrable Hamiltonian systems with simple singular points (topological aspects)*, Transl. Math. Monographs, vol. 176, Amer. Math. Soc., Providence, RI, 1998.
[6] V. I. Arnold, *Mathematical methods of classical mechanics*, Springer-Verlag, New York, 1989.
[7] L. M. Lerman and Ya. L. Umanskiĭ, *The structure of a Poisson action of $\mathbb{R}^2$ on four-dimensional symplectic manifolds*, Selecta Math. Soviet. (transl. of a Russian preprint of 1981) **6** (1987), 365–396.
[8] ———, *The structure of a Poisson action of $\mathbb{R}^2$ on a four-dimensional symplectic manifolds, 2*, Selecta Math. Soviet. (transl. of a Russian paper of 1982) **7** (1988), 39–48.
[9] ———, *Integrable Hamiltonian systems and Poisson actions*, Selecta Math. Soviet. (transl. of a Russian paper of 1984) **9** (1990), 59–67.

[10] A. T. Fomenko, *Topology of constant energy surfaces for integrable Hamiltonian systems and obstructions to integrability*, Izv. Akad. Nauk SSSR Ser. Mat. **50** (1986), 1276–1301; English transl. in Selecta Math. Soviet.

[11] A. T. Fomenko and H. Zieschang, *A topological invariant and criterion of equivalence for integrable Hamiltonian systems with two degrees of freedom*, Izv. Akad. Nauk SSSR Ser. Mat. **54** (1990), no. 3, 546–575; English transl., Math.-USSR-Izv. **36** (1991), no. 3, 567–596.

[12] L. M. Lerman and Ya. L. Umanskiĭ, *Classification of four-dimensional Hamiltonian systems and Poisson actions of* $\mathbb{R}^2$ *in an extended neighborhood of simple singular points*, I, Mat. Sb. **183** (1992), no. 12, 141–176; English transl., Russian Acad. Sci. Sb. Math. **77** (1994), no. 2, 511–542.

[13] ———, *Classification of four-dimensional Hamiltonian systems and Poisson actions of* $\mathbb{R}^2$ *in an extended neighborhood of simple singular points*, II, Mat. Sb. **184** (1993), no. 4, 105–138; English transl., Russian Acad. Sci. Sb. Math. **78** (1994), no. 2, 479–506.

[14] ———, *Classification of four-dimensional Hamiltonian systems and Poisson actions of* $\mathbb{R}^2$ *in an extended neighborhood of simple singular points*, III. Realization, Mat. Sb. **186** (1995), no. 10, 89–102; English transl. in Russian Acad. Sci. Sb. Math.

[15] A. V. Bolsinov, *Methods of the calculation of the Fomenko–Zieschang invariant*, Adv. Soviet Math., vol. 6, Amer. Math. Soc., Providence, RI, 1991, pp. 147–183.

[16] S. V. Matveev, *Integrable Hamiltonian systems with two degrees of freedom. Topological structure of saturated neighborhoods of focus-focus and saddle-saddle singular points*, Mat. Sb. **187** (1996), no. 4, 29–58; English transl. in Math. Russian Acad. Sci. Sb. Math.

[17] V. V. Kalashnikov, *Typical integrable Hamiltonian systems on a four-dimensional symplectic manifold*, Izv. Ross. Akad. Nauk Ser. Mat. **62** (1998), 49–74; English transl. in Russian Acad. Sci. Izv. Math.

[18] Nguyen Tien Zung, *A note on focus-focus singularities*, J. Geom Appl. (to appear)

[19] ———, Nguyen Tien Zung, *Symplectic topology of integrable Hamiltonian systems*, I: *Arnold–Liouville with singularities*, Compositio Math. **101** (1996), 179–215.

[20] J. E. Marsden and A. Weinstein, *Reduction of symplectic manifolds with symmetry*, Rep. Math. Phys. **5** (1974), 121–130.

[21] L. M. Lerman and Ya. L. Umanskiĭ, *Isoenergetic classification of integrable Hamiltonian systems in a neighborhood of a simple elliptic point*, Mat. Zametki **55** (1994), 88–97; English transl. in Math. Notes **55** (1994).

[22] V. I. Arnold, V. V. Kozlov, and A. I. Neishtadt, *Mathematical aspects of classical and celestial mechanics*, Encyclopaedia of Mathematical Sciences, vol. 3, Springer-Verlag, Berlin–Heidelberg–New York, 1988.

[23] N. K. Gavrilov, *On some bifurcations of codimension* 2 *equilibria for divergence-free vector fields*, Methods of Qualitative Theory of Differential Equation (E. A. Leontovich, ed.), Gor'kov. Gos. Univ., 1985, pp. 46–55.

[24] J. Vey, *Sur certains systèmes dynamiques séparables*, Amer. J. Math. **100** (1978), 591–614.

[25] H. Eliasson, *Normal forms for Hamiltonian systems with Poisson commuting integrals—elliptic case*, Comment. Math. Helv. **65** (1990), 4–35.

[26] A. T. Fomenko, *Bordism theory for integrable Hamiltonian nondegenerate systems with two degrees of freedom. A new topological invariant of multidimensional integrable systems*, Izv. Akad. Nauk SSSR Ser. Mat. **55** (1991), 747–779; English transl. in Math.-USSR-Izv.

[27] D. B. Fuks and V. A. Rokhlin, *Beginner's course in topology. Geometric chapters*, "Nauka", Moscow, 1977; English transl., Springer-Verlag, Berlin and New York, 1984.

[28] S. Smale, *Topology and mechanics*, Invent. Math. **10** (1970), no. 4, 305–331; **11** (1970), no. 1, 45–64.

Research Institute for Applied Mathematics and Cybernetics, 10 Ul'yanov Street., Nizhny Novgorod, 603005 Russia

*E-mail address*: `lerman@focus.nnov.ru`

Translated by THE AUTHOR

i 18870624